Open-Source Lab

Open-Source Lab

How to Build Your Own Hardware and Reduce Research Costs

Joshua M. Pearce

Department of Materials Science & Engineering,
Department of Electrical & Computer Engineering,
Michigan Technological University,
Houghton, MI, USA

ELSEVIER

AMSTERDAM ▪ BOSTON ▪ HEIDELBERG ▪ LONDON ▪ NEW YORK ▪ OXFORD
PARIS ▪ SAN DIEGO ▪ SAN FRANCISCO ▪ SYDNEY ▪ TOKYO

Elsevier
225 Wyman Street, Waltham, MA 02451, USA
The Boulevard, Langford Lane, Kidlington, Oxford OX5 1GB, UK
Radarweg 29, PO Box 211, 1000 AE Amsterdam, The Netherlands

Library of Congress Cataloging-in-Publication Data
Pearce, Joshua, author.
  Open-source lab : how to build your own hardware and reduce research costs / Joshua Pearce.
      pages cm
  Includes bibliographical references and index.
  1. Laboratories--Equipment and supplies. 2. Open source software. I. Title.
  Q185.P43 2013
  681'.750285--dc23
                            2013035658

British Library Cataloguing in Publication Data
A catalogue record for this book is available from the British Library

ISBN: 978-0-12-410462-4

Note: For color version of the figure, the reader is referred to the online version of this book.

For information on all **Elsevier** publications
visit our web site at store.eslevier.com

Working together
to grow libraries in
developing countries

www.elsevier.com • www.bookaid.org

Contents

FOREWORD vii
PREFACE.. ix
ACKNOWLEDGMENTS... xv

CHAPTER 1   Introduction to Open-Source Hardware for Science....... 1
CHAPTER 2   The Benefits of Sharing—Nice Guys and Girls do
Finish First 13
CHAPTER 3   Open Licensing—Advanced Sharing............................ 37
CHAPTER 4   Open-Source Microcontrollers for Science:
How to Use, Design Automated Equipment With
and Troubleshoot 59
CHAPTER 5   RepRap for Science—How to Use, Design, and
Troubleshoot the Self-Replicating 3-D Printer 95
CHAPTER 6   Digital Designs and Scientific Hardware.................... 165
CHAPTER 7   The Future of Open-Source Hardware and Science.... 255

INDEX.. 265

Foreword

At the heart of open-source hardware is the freedom of information. We are inherently free to open our devices as we wish and poke around. There are no laws inhibiting a consumer to unscrew their household items and take the lid off—though it most likely voids the warranty. But the freedom to repair, freedom to study, and freedom to understand needs to be accompanied with a freedom of accessible information: schematics, diagrams, code, and in short source files. Open-source hardware includes the previous freedoms and also grants the freedom to remix, remanufacture and resell an item, provided that the hardware remains open source.

History points to a multitude of repair manuals from cars to washing machines; patterns to follow from model airplanes to dresses; and recipes shared through friends and families for generations. Historically DIY (Do-It-Yourself) was not a fad but a way of life. Access to information coupled with a basic knowledge of tinkering has given consumers the power to fix more, waste less, and understand the physical world around them. But technologies are becoming more opaque, as their size gets smaller, making them more difficult to open and tinker. Historically, an important factor for understanding the physical world was that items were built on a human scale. Human scale is the one that humans can relate to and can visibly see with the naked eye. The scale of most objects previous to computing has been on the human scale. Items in our daily lives now include miniscule chip sets and tiny form factors that require schematics and code to diagnose, repair, or even understand. Perhaps no one understands this better than researchers themselves. With closed source and patented devices, there is no requirement to include source files so that people may understand the hardware. In many cases, steps are taken to obfuscate information from the consumer. In addition to documentation, many new inventions require special equipment and tools, such as laser cutters, PCR machines for DNA sequencing, environmental chambers and other lab equipment described in Pearce's work. These tools are beginning to see open-source versions so that consumers may build their own, often at a lower cost. Even more standard tools, such as tractors and

CNC machines are being open sourced so that others may have the benefit of access to these basic tools.

If history has favored open source, why are we entering a new movement of open-source hardware? Patents have become problematic to innovation. Basic building blocks of new technologies are being closed off with patents, causing further innovation to become increasingly expensive or halt altogether. While patenting the building blocks of technology may benefit one company, it fails to advance society. Today Intellectual Property can be sold as a good. The *idea* is the commodity rather than the physical object itself. Selling ideas rather than goods does not create a sustainable market for the common consumer. Patents were created to incentivize inventors and spur innovation in exchange for 20 years of exclusive rights in the form of a monopoly. The patentee had to submit a prototype and disclose how their innovation was created to the public. But the rules on patents have changed over time and there are many schools of thought that the patent system is broken and no longer reflects the reasons why the patent system was created in the first place. In today's patent system, prototypes are no longer required, money made from patents is going to lawyers rather than the inventor, and a 20 years monopoly is not a rational timeframe for the pace of technology in the digital era. Inventors are finding different incentives to innovate. The barriers and frustrations the patent system, has created are turning inventors toward a new alternative to patents: open-source hardware.

Open-source hardware creates products driven by capitalism rather than monopolies, an open environment for sharing information, and a powerful opportunity for companies and individuals to learn from each other. Open-source hardware is a growing movement with a lucrative business model. It has spread into many areas of innovation, as Pearce has done with his work in scientific hardware, others do in electronics, mechanical designs, space programs, farm equipment, fashion, and materials science to name a few. We are at a crucial point in the history of technology which will determine if we hoard information or share it with others; sell information or sell goods; educate with open documentation or let everyone reinvent the wheel for themselves.

—Alicia Gibb
Founder, Open Source Hardware Association

Preface

As the process of development that has succeeded in free and open-source software is now being applied to hardware, an opportunity has arisen to radically reduce the cost of experimental research in the sciences while improving the tools that we use. Specifically, the combination of open-source 3-D printing and open-source microcontrollers running on free software enables the development of powerful research tools at unprecedented low costs. In this book, these developments are illustrated with numerous examples to lay out a path for reinforcing and accelerating free and open-source scientific hardware development for the benefit of science and society. Wise scientists will join the open-source science hardware revolution to see the costs of their research equipment drop as their work becomes easier to replicate and cite.

Most scientists who do experimental work are familiar with and somewhat acclimated to the often extreme prices we pay for scientific equipment. For example, last year, I received a quote for a $1000 lab jack. A lab jack is not an overly special or sophisticated research tool; it simply moves things up and down like a jack for a car, only more precisely and the "things" it has to move are much smaller. That price for the application I was planning on using it for (moving millimeter-scale solar photovoltaic cells into a beam of light) was absurd, but as many researchers in academia know, the prices are effectively multiplied because of institutional overheads. Thus, at my institution for example, where we pay 71% overhead for industry-sponsored research, purchasing that lab jack would demand that I raise $1710 from sponsors! Historically, you, I and the rest of the scientific community had no choice—we had to buy proprietary tools to participate in state-of-the-art research or develop everything from scratch. Thus, we had to choose between paying exorbitant fees or investing a lot of our own time as even the simplest research tools like a lab jack are time-consuming to fabricate from scratch. No more! Now the combination of open-source microcontrollers and 3-D printers enables all of us to fabricate low-cost scientific equipment with far less-time investment than ever possible in the history of science. If you want the complete digital designs

FIGURE 1

An open-source lab jack. A lab jack is a height-adjustable platform ideal for mounting optomechanical subassemblies that require height adjustment. This lab jack was developed by the Michigan Tech Open Sustainability Technology Lab group primarily by Nick Anzalone and the gears were developed by Thingi-verse user: GregFrost. The OpenSCAD files (raw source code), STL files (for printing) and instructions can be found at http://www.thingiverse.com/thing:28298.

of a lab jack, you can print for a few dollars (Figure 1).[1] In fact, for $1710, you could buy yourself a nice 3-D printer and print hundreds of lab jacks and other high-value equipment for your friends, colleagues and family! Scientists have been catching on and the files for the lab jack have already been downloaded thousands of times.

Lower cost and less-time investment are actually only secondary benefits of using open-source hardware in your lab. The main advantages of the "open-source way" in science is customization and control. Rather than buy what is on the shelf or available from vendors online, you can create scientific instruments that meet your exact needs and specifications. This really is priceless (Figure 2). The ability to customize research tools is particularly helpful to those on the bleeding edge of science, which need custom, never-seen-before equipment to make the next great discovery. If the tools and software you use to run your experiments are open source, you and your lab group have complete control over your lab. If you hold the code, your lab will never be left empty handed (or stuck with extremely expensive paper weights) when commercial vendors go out of business, drop a product line, or lose key technical staff. Every research university in the world is sitting on millions of dollars of broken scientific equipment that is too expensive or time-consuming to repair. With open-source hardware, these problems largely evaporate. Similarly, if you only

[1]http://www.thingiverse.com/thing:28298

Commercial 6 position automated filter wheel: $2555.00
Shipping: $50.00
Availability: 5 weeks

Open-source 3-D printer: <$1000.00
An open-source automated 8 position filter wheel: ~$50.00
Printing and construction time: <1 day
The ability to customize scientific tools and make them yourself: Priceless

There's some things money can buy
And for everything else there is the RepRap.

FIGURE 2

The ability to customize high-quality research-grade scientific equipment is priceless, yet it often costs orders of magnitude less than conventional proprietary tools.

use open-source hardware rather than locked-down proprietary tools, your lab cannot be held for ransom by dishonest hardware vendors. It is perhaps important to note here that "open source" does not necessarily mean "free" as we will discuss in Chapter 1. Making open-source hardware following the process outlined in this book almost always results in much lower costs. However, the open-source scientific hardware movement is still developing and all the research tools you need may not have been developed yet. For scientific hardware that we do not fabricate in our own lab, we will pay a premium for open-source equipment and free software because of the access to the control or simply because it is superior (as discussed in Chapter 2). I am far from alone in thinking this way. One of the computers I am typing this book on is running a version of Linux[2] developed by Red Hat. Although Red Hat's software is freely available for download on the Internet, they make over US$1 billion per year essentially selling service to help people maintain it and optimize it for their applications.

---

[2]Linux is a free and open-source computer operating system, which comes in many varieties. The majority of the rest of the book was written on a laptop running Ubuntu Linux (http://www.ubuntu.com/) and my lab is currently making the transition to Debian (http://www.debian.org/).

In our lab, whenever possible, we want control over our own equipment, thus we use open-source hardware and software as much as possible. We now develop the majority of our tools in at least some way using the open-source paradigm. Now we have joined a virtuous cycle—as our research group shares our designs for research equipment with others, they make them better and share their improvements back with us. I should be clear here; this goes far beyond simple benevolent sharing or charity. **Our lab equipment quality improves because we actively share.** In this way, we all benefit.

For those working at academic, government and industry laboratories, this book is meant to be an introductory guide on how to join and take advantage of all the benefits of open-source hardware for science. Chapter 1 covers the basics of open source, a brief overview of the history and etiquette of open-source communities. The theoretical argument for why this method of technological development is superior to the standard models is laid out in Chapter 2, which describes why nice guys finish first both in research and at the industrial scale. Chapter 3 covers the nitty-gritty of open-source licensing. Chapters 4 and 5 describe the two most useful tools for fabricating open-source scientific equipment, the Arduino microcontroller and RepRap 3-D printer, respectively. Chapter 6 is the heart of the book and describes digital designs for open-source scientific hardware in physics, engineering, biology, environmental science and chemistry. Finally, in Chapter 7, we take a peek at the future and possible ramifications of large-scale adoption of open-source hardware and software for science.

A note to readers: This book is not necessarily the kind of text you read from cover to cover. If you are already "at one with the force of open source" and just want to get your hands dirty—start building by skipping to Chapter 4. Similarly, only a few of the examples in Chapter 6 will be relevant to your work— no need to learn about PCR if you are a physicist researching new nanoscale solid-state detector technologies—just go right to the tools that will be most helpful to you… share and enjoy.

1.1. STANDARD DISCLAIMER

Knowledge and best practice in this field are constantly changing. As new research and experience broaden our understanding, changes in research methods, professional practices, or safety may become necessary.

Practitioners and researchers must always rely on their own experience and knowledge in evaluating and using any information, methods, equipment, compounds, or experiments described herein. In using such information or methods, they should be mindful of their own safety and the safety of others, including parties for whom they have a professional responsibility.

To the fullest extent of the law, neither the Publisher nor the author, contributors, or editors assume any liability for any injury and/or damage to persons or property as a matter of products liability, negligence or otherwise, or from any use or operation of any methods, products, instructions, or ideas contained in the material herein.

Joshua M. Pearce, Ph.D.

Associate Professor
Department of Materials Science & Engineering Department of Electrical &
Computer Engineering Michigan
Technological University Houghton, MI USA

Acknowledgments

First and foremost, and on a personal note, I would like to thank my wonderful wife, Jen, for putting up with the long hours of writing this text, turning various parts of our house into electronics, 3-D printing and scientific labs as the need arose, her support and also for reading and critiquing the manuscript.

I would like to thank my children, Emily and Jerome, who actually helped with some of the making of my first 3-D printer and with various experiments throughout the years.

I would also like to thank the rest of my family: Mom and Dad, Solomon, Mary Rachel and Elijah for their support and encouragement. A special thanks to Mary Rachel for giving me a copy of *Makers* by Cory Doctorow, which laid some of the ground work for this book.

I would like to thank Elsevier for having the foresight and courage for both publishing this book and also ensuring that it is maintained in the open-source ethos by making it available to the widest possible scientific audience. For this, I especially thank Beth Campbell, Jill Cetel, Paula Callaghan, Cathleen Sether, and Laura Colantoni for making this book a reality.

This book was truly a massive international and asynchronous collaboration that extends back years and contains the genius and insights of people I have worked closely with, but also many whom I have never met (or may only know of through their esoteric Internet handles).

First, from the past and present members of my own research group at Michigan Tech—the Michigan Tech Open Sustainability Technology Lab, I would like to thank for fruitful collaboration: Nick Anzalone, Megan Kreiger, Chenlong Zhang, Ben Wittbrodt, Allie Glover, Brennan Tymrak, Meredith Mulder, Ankit Vora, John Laureto, Joseph Rozario, Jephias Gwamuri, Alicia Steele, Thad Waterman, and Paulo Seixas Epifani Veloso. Special thanks goes to Rodrigo Periera Faria, Bas Wijnen and Jerry Anzalone for collaboration and contributions to major sections of this book and for reading drafts of portions of the manuscript. I would also like to thank other Michigan Tech collaborators and supporters: Doug Oppliger, John Irwin, Allison Hein, and the support and leadership of both of my department heads, Steve Kampe in Materials Science and Engineering and Dan Fuhrmann in Electrical and Computer Engineering.

In addition, my former students at Queen's University helped me get started with the wondrous RepRap and on the train of thought that led to this book: Christine Morris Blair, Kristen Laciak, Steven Keating, Christian Baechler, Matthew DeVuono, Amir Nosrat, Ivana Zelenika, and Rob Andrews.

I would also like to acknowledge the funders, supporters and corporate sponsors for some of this and related work, including the McArthur Internship at Michigan Tech, the National Science Foundation, Natural Sciences and Engineering Research Council of Canada, Superior Ideas (and supporters through it), the Appropedia Foundation, Re:3D, Tech for Trade, Ocean Optics, Type A Machines, MatterHackers, Ultimachine, and the Square One Education Network.

I would like to acknowledge helpful discussions about enabling innovation from Scott Albritton, Gabriel Grant and Garrett Steed. For their ongoing support, I thank the Appropedia community and, in particular, Lonny Grafman and Chris Watkins. A giant thank you goes to the entire GNU/Linux community for really showing us what is possible when we all work together and providing us with the free software we rely on. I would also like to thank the Arduino founders, Massimo Banzi and David Cuartielles, and all their collaborators for making the control of scientific equipment easy and fun. The entire world owes a great thanks to Adrian Bowyer and his many collaborators on making the RepRap project into the incredible success that it is. I would like to thank all the fantastic open-source hardware individuals, groups and companies that keep enabling us to reach higher. I want to thank the Open-Source Hardware Association and all their members, in particular, Alicia Gibb and Catarina Mota. I want to thank all those who have shared their brilliance with the world and help make my research a success because of their contributions in the scientific and engineering literature, as well as Github, Thingiverse, and Appropedia users. A special thanks to everyone who provided examples that are cited or shown in the book. Finally, I would like to thank the growing number of makers in the burgeoning maker community that inspire and teach us all.

DISCLAIMER

Although many people contributed to the contents of this book, all errors and omissions are mine alone. The technologies described in this book are constantly changing, and while every effort was made to ensure accuracy of this work, it is always best to go directly to the sources for the most up-to-date information on the various open-source hardware projects described.[1]

Finally, if any of the hardware is not good enough for you or your lab, remember it is free so quit whining and make it better!

---

[1]Wherever possible hyper links are shown in the footnotes and will be enabled on the digital version of this book.

CHAPTER 1

Introduction to Open-Source Hardware for Science

1.1. INTRODUCTION

By any standard, the process of development and licensing for free and open-source software, which is discussed in Chapters 2 and 3, has been a success. Because of this success, the method is now being applied to hardware. Thus, an opportunity has arisen to radically reduce the cost of experimental research in the sciences [1]. This opportunity has the potential to reduce your research costs and make your laboratory more productive—while at the same time vastly expanding the scientific user base. Specifically, this book focuses on the combination of open-source microcontrollers covered in Chapter 4 and open-source 3-D printing reviewed in Chapter 5. These two tools running on free open-source software enable the development of powerful research tools at unprecedented low costs. Chapter 6 provides several detailed examples of these tools for a wide range of science and engineering disciplines. Then, in Chapter 7, these developments and their likely trajectories in the future are explored and illustrated with numerous examples to lay out a path of mutually reinforcing and accelerating free and open-source scientific hardware development for the benefit of science.

1.2. WHAT IS OPEN SOURCE?

The term "open source" emerged during a strategy session between several hackers[1] of the early open software movement [2]. Free and open-source software (F/OSS, FOSS) or free/libre/open-source software (FLOSS) is a software that is both a free software and an open source. *FOSS* is a computer software that is available in source code (open source) form and that can be used, studied, copied, modified, and redistributed without restriction, or with restrictions that only ensure that further recipients have the same rights under which it was obtained (free or libre). *Free software, software libre or libre software* is software that can be used, studied, and modified without restriction, and which can be copied and redistributed

---

[1]Although the term "hacker" is often associated with illicit activity in public discourse, it actually refers to a computer programmer who develops free and open-source software.

Open-Source Lab. http://dx.doi.org/10.1016/B978-0-12-410462-4.00001-9

in modified or unmodified form either without restriction or with restrictions that only ensure that further recipients have the same rights under which it was obtained and that manufacturers of consumer products incorporating free software provide the software as source code. The word "free" in the term refers to freedom (liberty) and is not necessarily related to monetary cost. Although FOSS is often available without charge, it is not bound to such a restriction.

GNU/Linux (normally called just Linux), perhaps, is the best example of an open-source project created by the many. Linus Torvalds originally developed Linux in 1994 as a little hacking project to replace a program called Minix, which was a teaching tool in computer science courses. He released the source code to everyone using the Internet. The hacker and the software developer community at-large were immediately enthralled by Linux and began contributing improvements to the source code. The program went from being a little side project to a full PC-based operating system to which more than 3000 developers distributed over 90 countries on five continents contributed. In the first few years of its development, more than 15,000 people submitted code or feedback to the Linux community. Linux went from consisting of a few hundred lines of code to several millions of lines of code. Despite this rapid growth and large developer community, the reliability and quality of the operating system were ranked very highly [3]. These observations provide support of Raymond's claim that "given enough eyeballs, all bugs are shallow" [4], meaning that problems become easier to solve with more collaborators. This is the fundamental strength of the open-source paradigm. We are all smarter than any one of us. Although the basics of open source remain the same since the start of the movement, the current definition of open source has been expanded to include such criteria as free redistribution rights and no discrimination against people/groups in accessing source code [5].

As much of the Internet now relies on FOSS, we all use it every day. Even if your laptop is not running a version of Linux like the one I am typing on now (System 76 running Ubuntu[2]), the back ends of the world's most popular websites are all run on open-source software (e.g. Google, Facebook, etc.). FOSS is not only just as good as commercial software—it is often superior. If an open-source software such as GNU/Linux is compared head-to-head against Microsoft's centralized and closed system of software development, a perhaps surprising result surfaces. A neutral technical assessment finds that open-source

---

[2]Ubuntu is a computer operating system based on the Debian Linux distribution and distributed as free and open-source software, using its own desktop environment. It is the most popular version of Linux with over 20 million people using it as of 2012. Ubuntu is a South African ethical ideology focusing on people's allegiances and relations with each other. Rough translations of the principle of ubuntu is "humanity towards others" or "the belief in a universal bond of sharing that connects all humanity". Ubuntu, the operating system can be downloaded for free from http://www.ubuntu.com/ and Debian from http://www.debian.org/.

software, developed in the early days mostly by unpaid volunteers, is often of superior quality to the software developed by one of the most powerful companies in the history of the world employing unquestionably extremely intelligent people [6]. This remarkable result stands against conventional wisdom that would argue the profit motive and market forces would enable Microsoft to develop superior software to any random group of volunteers. Microsoft is a large company, with annual revenue of over US$40 billion, yet many of its products suffer from technical drawbacks that include bloat, lack of reliability, and security holes. Microsoft remains dominant in the PC market largely because of inertia, but Linux eats up an ever larger market share (particularly in servers), because open source is simply more efficient and adaptable than closed, hierarchical systems [7]. This is due, historically at least, in a large part because a lot more people collaborate on Linux than on Microsoft products. Whereas Microsoft might utilize a few thousand programmers and software engineers to debug their code, the Linux community has access to hundreds of thousands of programmers debugging, rewriting, fixing, making suggestions and submitting code.[3] Eric Raymond in the *Cathedral and the Bazaar* argued that open source was a fundamentally new way to create and design technology that relied on the "eyeballs" of the many instead of the minds of the few. Using the cathedral and bazaar analogy, Raymond claimed that open source, drawing from the rich, nonhierarchical, gift-based culture of hackers, was similar to the bazaar where everyone could access and contribute equally in a participatory manner. This type of mass-scale collaboration is driving the success of Web 2.0 applications that emphasize online collaboration and sharing among users (examples include social networking sites and wikis such as Wikipedia). Wikipedia is a particularly good example because both the software is open source, as is the development method of the content, which has proven both accurate [8] and far superior in terms of both total and up-to-date coverage compared to conventional for-profit encyclopedia generation as summarized in Figure 1.1 [9].

This superior software development method has even developed something of an actual "movement". The FOSS movement emerged as a fundamentally new, decentralized, participatory and transparent system to develop software in contrast to the closed box, top-down and secretive standard commercial approach [4,10–12]. FOSS provides (1) an alternative to expensive and proprietary systems, (2) a reduction in research and development costs, (3) a viable alternative to the linear hierarchical structure used to design any type of technology-based products and (4) the efficiency of collaboration, demand-driven innovation and the power of the Internet to provide for a global collective

---

[3]In fact, even Microsoft is now embracing some components of open-source software development. A Microsoft representative has recently stated that both SQL Server and the Windows Azure teams are committed to the Hadoop open-source platform for the long term [30].

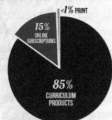

FIGURE 1.1

Wikipedia infographic by Statista.

good. Due to this tremendous success of FOSS development, the concept has spread to areas such as education [13,14], appropriate technology for sustainable development (called open-source appropriate technology) [15–18], science [19], nanotechnology [20–22] and medicine [23,24]. Both academic and nonacademic scientists are accustomed to this line of thinking as both historical knowledge sharing and the Internet enabled new era of networked science have demonstrated the enormous power of working together [25–27].

1.3. FREE AND OPEN-SOURCE HARDWARE

These open and collaborative principles of licensing FOSS are easily transferred to scientific hardware designs [1]. Thus, *free and open-source hardware* (*FOSH*) is a hardware whose design is made publicly available so that anyone can study, modify, distribute, make, and sell the design or hardware based on that design. The most successful enabling open-source hardware project is the Arduino electronic prototyping platform[4], which we will investigate in detail in Chapter 4. The $20–60 Arduino is a powerful, yet easy-to-learn microcontroller that can be used to run a burgeoning list of scientific apparatuses directly including the already developed Polar Bear (environmental chamber—detailed in Chapter 4), Arduino Geiger (radiation detector)[5], pHduino (pH meter)[6], Xoscillo (Oscilloscope)[7], and OpenPCR (DNA analysis)[8]. However, one of the Arduino's most impressive technological evolution-enabling applications is with 3-D printing.

Using an Arduino as the brain, 3-D printers capable of additive manufacturing or additive layer manufacturing from a number of materials including polymers, ceramics and metals have been developed. The most popular of these 3-D printers is the RepRap, named because it is a self-replicating rapid prototyping machine.[9] Currently, the RepRap (Figure 1.2), which uses fused-filament fabrication of complex 3-D objects, can fabricate approximately 50% of its own parts and can be made for under $1000 [28]. The version we will explore in Chapter 5 can be built for about $500 and assembled in a weekend. This ability to inexpensively and freely self-replicate has resulted in an explosion of both RepRap users and design improvements [28]. RepRaps are used to print many kinds of objects from toys to household items, but one application where their transformative power is most promising is in significantly reducing

---

[4]http://www.arduino.cc/.
[5]http://www.cooking-hacks.com/index.php/documentation/tutorials/geiger-counter-arduino-radiation-sensor-board.
[6]http://phduino.blogspot.com/.
[7]http://code.google.com/p/xoscillo/.
[8]http://openpcr.org/.
[9]If you have never seen a RepRap work, put this book down and go to http://reprap.org/—click on the video and then come back.

FIGURE 1.2

A self-replicating rapid prototyper known as the RepRap can 3-D print about 50% of its own parts (including the component being fabricated) and numerous scientific tools and components. All the red, white and gray components were printed on other RepRaps to make this highly customized version of the Mendel Prusa developed by Jerry Anzalone.

experimental research costs. As many scientists with access to RepRaps have found, it is less expensive to design and print research tools, and a number of simple designs have begun to flourish in Thingiverse[10], which is a free and open repository for digital designs for real physical objects. These include single-component prints such as parametric cuvette/vial racks as shown in Figure 1.3. Three-dimensional printers have also been used to print an entirely new class of reactionware for customizing chemical reactions [29].

Combination devices have also been developed where a 3-D print is coupled to an existing hardware tool (such as the portable cell lysis device for DNA extraction shown in Figure 1.4), which is a 3-D printable adapter that converts a Craftsman automatic hammer into a bead grinder for use with DNA extraction.

Similarly, the DremelFuge chuck shown in Figure 1.5 is a 3-D printable rotor for centrifuging standard microcentrifuge tubes and mini-prep columns powered by a Dremel drill.

---

[10]http://www.thingiverse.com/.

FIGURE 1.3
Parametric cuvette and vial rack. *Source: Open sourced by Emanresu* (http://www.thingiverse.com/thing:25080).

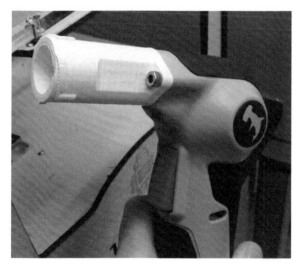

FIGURE 1.4
A 3-D printable adapter to turn a craftsman automatic hammer into a bead grinder for use with DNA extraction. *Source: Open sourced by Russell Neches at U.C. Davis* (http://www.thingiverse.com/thing:18289).

These combination devices can radically reduce research costs. For example, the DremelFuge can be used in the lab or the field as an extremely inexpensive centrifuge (price: <$50, primarily for the Dremel drill—compared to commercial centrifuges systems, which cost a few hundred dollars).

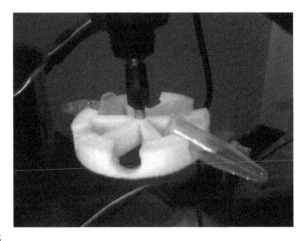

FIGURE 1.5
The DremelFuge, a 3-D printable rotor for centrifuging standard microcentrifuge tubes and mini-prep columns powered by a Dremel drill. *Source: Designed and open sourced by Cathal Garvey* (http://www. thingiverse.com/thing:1483).

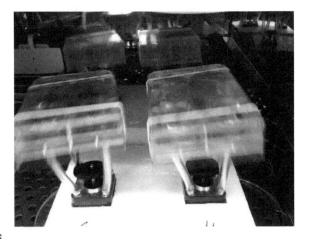

FIGURE 1.6
Open-source orbital shaker using 3-D printable components. *Source: Open sourced by Jordan Miller at the University of Pennsylvania* (http://www.thingiverse.com/thing:5045).

The most aggressive savings can come from coupling Arduino controls to 3-D prints to make open-source scientific hardware. Consider the Arduino-controlled open-source orbital shaker in Figure 1.6, used for mammalian cell and tissue culture and bench-top science. The <$200 open-source orbital shaker fits inside a standard 37°C/5% CO_2 cell incubator and replaces commercial versions that

FIGURE 1.7

Parametric automated filter wheel changer. Inset shows detail of filter locking. *Source: Open sourced by the author, Rodrigo Faria and Nick Anzalone of the Michigan Technological University Open Sustainability Technology Group* (http://www.thingiverse.com/thing:26553).

start over $1000—a factor of 5X savings. As the scientific tools that are open-sourced gain complexity, the cost differential becomes even more severe.

For example, as seen in Figure 1.7, it is now possible to make a <$50 customizable automated filter wheel that replaces a $2500 commercial version (or a factor of 50X savings). The filter wheel changer in Figure 1.7 uses an open-source Arduino microcontroller to operate. All of the code to make and use it is open-source, including the design files, which are scripted in OpenSCAD, itself an open-source software tool. It was written carefully in OpenSCAD to be parametric, so other scientists can easily adjust the number or size of filters for their specific applications. A few months after it was published, the designs had been downloaded over 750 times, and presumably in use in labs throughout the world.

As additional research groups begin to freely share the designs of their own laboratory hardware, not only can everyone in the greater scientific community enjoy those same discounts on equipment, but also following the FOSS approach, the equipment will continue to evolve to be even better in the open-source scientific design community. In addition, research costs will also be depressed even when scientists choose a commercial version of a tool because of the price pressure from the open-source community.

We are on the verge of a new era where low-cost scientific equipment puts increasingly sophisticated tools into the hands of the public and amateur scientists, while driving down the costs of research tools at our most prestigious laboratories.

REFERENCES

[1] Pearce Joshua M. Building research equipment with free, open-source hardware. Science 2012;337(6100):1303–4.

[2] Bretthauer DW. Open source software: a history. Inf Technol Libr 2002;21(1):3–10.

[3] Moon JY, Sproull L. Essence of distributed work: the case of the Linux kernel. First Monday 2000;5(11); Retrieved 25.01.13, from: http://firstmonday.org/htbin/cgiwrap/bin/ojs/index.php/fm/article/view/801/710.

[4] Raymond ES. The cathedral and the bazaar. First Monday 1988;3(3); Retrieved 25.01.13, from: http://firstmonday.org/htbin/cgiwrap/bin/ojs/index.php/fm/article/view/578/499.

[5] The Open Source Definition (Version 1.9). Open source initiative. Retrieved 25.01.13, from: http://opensource.org/docs/definition.html.

[6] Bonaccorsi A, Rossi C. Why open source software can succeed. Res Policy 2003;32:1243–58.

[7] Kogut B, Metiu A. Open-source software development and distributed innovation. Oxford Rev Econ Policy 2001;17(2):248–64.

[8] Giles J. Internet encyclopaedias go head to head. Nature 2005;438:900–1.

[9] Silverman M. Encyclopedia Britannica vs Wikipedia. [Infographic] [WWW Document], Mashable 2012. Retrieved 25.02.13, from: http://mashable.com/2012/03/16/encyclopedia-britannica-wikipedia-infographic/.

[10] Mockus A, Fielding RT, Herbsleb JD. Two cases studies of open source software development: Apache and Mozilla. ACM Trans Software Eng Methodol 2002;11(3):309–46.

[11] Deek FP, McHugh JAM. Open source: technology and policy. New York, NY: Cambridge University Press; 2007.

[12] Weber S. The success of open source. Cambridge, MA: Harvard University Press; 2004.

[13] Pearce JM. Appropedia as a tool for service learning in sustainable development. J Educ Sustainable Dev 2009;3(1):47–55.

[14] Christian W, Esquembre F, Barbato L. Open source physics. Science 2011;334:1077–8.

[15] Pearce JM, Mushtaq U. Overcoming technical constraints for obtaining sustainable development with open source appropriate technology, science and technology for humanity (TIC-STH). 2009 IEEE Toronto International Conference; 2009, pp. 814–820.

[16] Buitenhuis AJ, Zelenika I, Pearce JM. Open design-based strategies to enhance appropriate technology development. Proceedings of the fourteenth annual national collegiate inventors and innovators alliance conference. 2010; 25–27: P. 1–2, March, 2010.

[17] Zelenika I, Pearce JM. Barriers to appropriate technology growth in sustainable development. J Sustainable Dev 2011;4(6):12–22.

[18] Pearce JM. The case for open source appropriate technology. Environ, Dev Sustainability 2012;14:425–31.

[19] Stokstad E. Open-source ecology takes root across the world. Science 2011;334:308–9.

[20] Bruns B. Open sourcing nanotechnology research and development: issues and opportunities. Nanotechnology 2001;12:198–210.

[21] Mushtaq U, Pearce JM. Open source appropriate nanotechnology. In: Maclurcan D, Radywyl N, editors. Nanotechnology and global sustainability. New York, NY: CRC Press; 2012. p. 191–213.

[22] Pearce JM. Make nanotechnology research open-source. Nature 2012;491:519–21.

[23] Lang T. Advancing global health research through digital technology and sharing data. Science 2011;331:714–7.

[24] Meister S, Plouffe DM, Kuhen KL, Bonamy GMC, Wu T, Barnes SW, et al. Imaging of plasmodium liver stages to drive next-generation antimalarial drug discovery. Science 2011;334:1372–7.

[25] Woelfle M, Olliaro P, Todd MH. Open science is a research accelerator. Nat Chem 2011;3(10):745–8.

[26] Nielsen M. Reinventing discovery: the new era of networked science. Princeton University Press; 2011.

[27] Pearce JM. Open source research in sustainability. Sustainability: J Rec 2012;5(4):238–43.

[28] Jones R, Haufe P, Sells E, Iravani P, Olliver V, Palmer C, et al. RepRap – the replicating rapid prototyper. Robotica 2011;29:177–91.

[29] Symes MD, Kitson PJ, Yan J, Richmond CJ, Cooper GJT, Bowman RW, et al. Integrated 3D-printed reactionware for chemical synthesis and analysis. Nat Chem 2012;4:349–54.

[30] Metz C. Microsoft embraces elephant of open source. Wired 2011. Retrieved 25.02.13, from: http://www.wired.com/wiredenterprise/2011/10/microsoft-and-hadoop/.

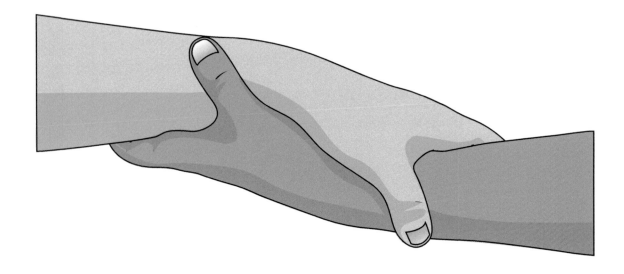

CHAPTER 2

The Benefits of Sharing—Nice Guys and Girls do Finish First

2.1. ADVANTAGES OF AGGRESSIVE SHARING FOR THE ACADEMIC

The primary purpose of universities is to spread knowledge, yet ironically, research faculty members are often encouraged to restrict information sharing. Today at many universities, there is enormous pressure to avoid having so-called intellectual property (IP) "scooped" by patenting and/or commercializing research to prevent replication without the academic's employer benefiting financially [1]. As the well-documented influence of corporate thought on universities has spread [2], this intellectual monopoly[1] view of research has now even infiltrated the academic literature. As can be seen in many fields, normally the experimental section of journal articles is the shortest and most opaque section of a manuscript. This makes it difficult to replicate our peers' experiments and may even potentially "threaten the foundation of scientific discourse" as Gelman argues [3]. Even the details that are provided are sometimes delayed. For example, many universities encourage holding back key information, which is not released until provisional patents have been filed, slowing the scientific enterprise down. In addition, most experimental protocols that are standards are not open access and many require substantial fees to even view (e.g. ASME standards). In addition, we are all familiar with the retarding force on further and faster scientific development from the lack of universal open access to the literature (discussed in detail below). All these trends hurt scientific communication and directly hamper innovation and progress.

Although this is common knowledge in academia, academics have learned and are accustomed to the system. Thus, it may be tempting to some research labs to learn all they can about making less expensive and more customizable research equipment using the method and sources described in the next several chapters, make the equipment they need and get on with their research

---

[1]See *Against Intellectual Monopoly* for a detailed evisceration of the justification for the existence of intellectual property (copyright or patents) [56]. Consistent with their book's message, the entire text can be read at http://www.micheleboldrin.com/research/aim.html or at http://levine.sscnet.ucla.edu/general/intellectual/againstfinal.htm.

Open-Source Lab. http://dx.doi.org/10.1016/B978-0-12-410462-4.00002-0

following the standard relatively closed model. However, if you do this, you will miss out on the primary benefits of using the open-source method. There are five extremely valuable benefits to aggressively sharing your own protocols, methods, and hardware back into the open-source scientific community:

1. massive peer-review in the development of background material and experimental design, which leads to
2. improved experimental design and hardware design (often for radically lower costs), which provides higher performance equipment,
3. increased visibility, citations and public relations, which leads to
4. increased funding opportunities and improved student recruitment, and
5. improved student research-related training and education.

In the sections below, we explore each benefit in detail.

In order to take advantage of these benefits, it is important to understand the ethic behind open-source philosophy. This ethic was developed first by computer programmers working on free and open-source software (FOSS). Through principles of sharing and open access, open-source development treats users as developers by encouraging contribution, recognizing good work through peer approval, and propagating superior code [4]. This philosophy of the open-source movement is described by Levy as the "hacker ethic" with the following general principles [5]:

1. Sharing
2. Openness
3. Decentralization
4. Free access
5. World improvement.

This philosophy is enabled by the gift culture of open source, in which recognition of an individual is determined by the amount of knowledge given away [6]. In this system, *the richer you are, the more you give; the more valuable the gift, the more respect you gain*. This type of philosophy should be familiar to many in academia, which has also historically followed a gift culture, which rewards contributors through a process of peer review. We share our brilliance at professional conferences and through articles in the literature, and gain respect. The more you give away and the more valuable it is scientifically, the better is your career. In the academy, knowledge is our currency not money. Can we do more? We will explore ways to more aggressively share and harvest the rewards from each benefit of using open-source approaches to research in the following subsections.

2.1.1. Pre peer-review in the development of background material and experimental design

Between 1998 and 2013, my research groups have experimented with progressively more aggressive sharing of research ideas *before* they are put into practice. The degree to which your laboratory can commit to openness will depend on the field and your institution's rules. However, the evidence now is fairly well established that it would benefit every field to be open. This was perhaps recently explained most clearly by Nobel Prize winners Eric Maskin and his coauthor James Bessen when they found that when discoveries are "sequential" (so that each successive invention builds in an essential way on its predecessors), patent protection discourages innovation [7]. That is worth repeating—the standard IP regime that was ostensibly established to foster innovation actually retards it. Bessen and Maskin found that society and even inventors themselves may be better off without such protection and even an inventor's prospective profit may actually be enhanced by competition and imitation [7]. In the case of the sciences, where profit is not normally considered our sole motivation and more particularly, experimental equipment design, the topic of this book, it is clear that it should always be open in order for science to progress as rapidly as possible. For individual scientists and engineers following an open-source methodology, the advantages listed above can be acquired.

The experiments in my lab were a start at applying the lesson learned from the natural experiment of software patents analyzed by Bessen and Maskin to research. Overall these experiments realizing the hacker ethic for open-source research were successful as we observed an increase in quality and quantity of both research and applications [8]. This success was driven both by the web 2.0 platforms used in the experiments and by simply providing open access to much of the content of our work. First, there were several advantages we saw as academics by using an open wiki.[2] Wikis like Wikipedia, which most everyone is familiar with, are web pages that are easy to edit with a relatively shallow learning curve (e.g. a typical student can master the basics of wiki markup in under 30 min). Multiple members of our research team or collaborators outside our university are able to edit pages (e.g. a literature search or methodology description) at the same time or from different locations. Utilizing the wiki keeps track of the status of an individual or a team of researchers and enables easy updates to work from prior years. It also easily enables multiple members of the group to collaborate asynchronously. These advantages are true for any wiki (even those that are password protected within a single institution), but what makes an open wiki useful is that it facilitates others outside of the research group to assist your research—to participate in the open-source

---

[2]As most of the research performed by my group is related to applied sustainability, we use Appropedia (http://www.appropedia.org/), which is the primary site for collaborative solutions in sustainability, poverty reduction and international development on the Internet. There are equivalent types of wikis developed in other disciplines or research focuses that may be appropriate for your group. I recommend using the closest fit with largest user base you can find as such a wiki will provide the greatest feedback and support for your research.

way, the same way teams of open-source software developers collaborate. For example, we often see people from outside our group help us make our literature reviews[3] more complete, accurate and up-to-date. This saves us time and directly improves our work. Other examples of assistance from wiki users not affiliated with our research group include (1) giving helpful comments on the discussion tab of group pages, (2) fixing grammar/spelling errors and typos, (3) making improvements to algorithms and electronic tools, (4) correcting mistakes or improving our 3-D printable designs, and (5) listing our work on other sites, categories within the wiki and hyperlinking either to or within work that a group has made, which adds to the value and accessibility of the work. In all these cases, by sharing our work as we do it, others help us to improve it free of charge. Most importantly, we often see others submitting improvements to our methods and experimental apparatus design. This improves our group's research quantitatively.

2.1.2.   Improved experimental design and hardware design

These open research-based experiments involved openly sharing our methods, protocols, and experimental designs on the wiki. The first advantage is intrinsic; by stressing to the students that these experimental designs will be web-searchable for all the time, I found that graduate students are more careful about their experimental designs as they are being shared as a quasi-publication. This type of sharing ensured high quality in student experimental and simulation designs and reduced sloppiness during experiments. These intrinsic benefits were dwarfed, however, by external support because of sharing. Many times academics, industry and government scientists and engineers from all over the world have improved our experimental designs. These helpers external to the group have recommended new software or ways to use existing software we were already using in the group. Open-source software coders have improved programs, device drivers, and firmware to meet our needs[4], and many others have made specific recommendations and provided helpful advice on everything from component 3-D designs in our experimental rigs to electronics in other experimental setups.[5] In some cases, external supporters helped us correct errors and oversights in our write-ups before we started

---

[3]For live versions of the literature reviews our group maintains and are working on now see http://www.appropedia.org/Category:MOST_literature_reviews.

[4]For a list of useful open-source software for scientists and engineers, go to http://www.appropedia.org/Open_source_engineering_software. If a program you have used or developed is not on the list—please add it—click edit in the appropriate section.

[5]This occurs for every aspect of our experimental work on Appropedia, but also specifically on 3-D designs on Thingiverse, where we upload our printable designs. http://www.thingiverse.com/ is a database of open-source designs of physical objects—primarily focusing on 3-D printable designs.

nonoptimized experiments, which saved us enormous quantities of resources (time and money) by avoiding the need to repeat poorly optimized experiments. These benefits all came from massive peer-review and the fact that we actively shared. For example, many of our examples of using open-source hardware and software have been viewed over 10,000 times. There is a lot of eyeballs looking for potential mistakes and better ways of running experiments!

We have also benefited directly from becoming involved in open-source hardware development started by others. For example, many years ago, David Denkenberger and I worked on a project to simulate the lowest cost method of using solar energy to provide clean and safe drinking water in the developing world [9]. This theoretical work appeared promising, and I built prototypes while away from the university (to both the horror and amusement of my new wife the first prototypes were tested in our small bathroom, with a "solar simulator" hanging from the shower rod and my sophisticated electronics seated on the 'throne'). These prototypes used conventional (and expensive) heat exchangers and the whole apparatus functioned well enough to provide drinking water easily for a typical family. However, to meet the low-cost demands, we needed a highly effective, extremely low-cost heat exchanger, which simply did not exist in the market. Years passed, but we finally developed one using a thin, polymer-based expanded microchannel design for the heat exchanger [10]. The idea behind this type of heat exchanger is that we make up for the low thermal conductivity of the polymer by using extremely thin layers of plastic and thin channels. After enormous effort and absurd prototyping costs, we got one to work with high efficiency using standard, black garbage bags as the material! This was pretty exciting, but to "rent" polymer laser welding time at the rates that we made the first prototype (>\$1000/sample) to do any kind of serious research and development work would be cost prohibitive. At the same time, several open-source laser cutter rigs were being developed. We derived our system from Thingiverse user Peter's (Peter Jansen, a graduate student at McMaster University in Canada) design of a 3D-printable Laser Cutter.[6] Taking this design, mounting an appropriate fiber laser, customizing a few parts of the rig with 3-D printed parts (to be discussed in Chapter 5), and writing our own Arduino-based control code (Chapter 4), we had a polymer laser welder (Figure 2.1) for the cost of a few samples.[7] The development time for such a system with both the start and continued support from the open-source hardware community (made possible by our resharing of the designs) was a tiny fraction of what we would have needed to invest to develop the tool ourselves.

---

[6]See http://www.thingiverse.com/thing:11653.
[7]For complete designs, see http://www.appropedia.org/Open_source_laser_system_for_polymeric_ welding and 3-D printable variations on the parts http://www.thingiverse.com/thing:28078. Operating instructions: http://www.appropedia.org/Laser_welding_protocol:_MOST.

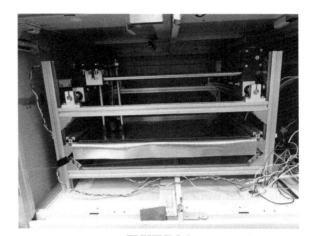

FIGURE 2.1

Open-source polymer laser welding system for microchannel heat exchanger fabrication.

Now we literally save several thousand dollars a day every day we make multiple sample heat exchangers.

Finally, it should be noted that by making not only our experimental designs open but also our experimental apparatuses, we have made it extremely easy for other groups to replicate and build on our work. Other groups can quickly build copies of our experimental apparatuses and begin collaborating to push our research work forward. This sharing again directly increases the impact our work has in various science and engineering subcommunities that we work within and can assist in increasing citations, a topic we will cover in detail in the next section.

2.1.3.   Increased visibility, citations and public relations

The open wiki that my group uses is Appropedia, which is currently the largest wiki dedicated to appropriate technology and sustainability, thus giving us widespread readership and accessibility. Using a page-view counting tool,[8] it is easy to see that our groups' work had been accessed over 1 million times and that it is accessed hundreds of thousands of times each year. This is an incredible visibility for academic work. This visibility of our various innovations and studies often attract media attention, which increases positive public relations for my university. Depending on your particular university, this may or may not be that important in tenure and promotion. However, no matter what university you are associated with, your citations will matter.

---

[8]http://pequals.com/appro_count.php.

An open-source research approach can directly improve your citations as discussed below.

When we have completed a project or paper, we place summaries of work on Appropedia, which often include some of the most useful data along with DOI links to the original article and ideally to an open-access version when tolerated from the publisher–author's agreement.[9] This practice unquestionably increases the number of people reading a given article. This not only helps our work have greater impact and accessibility in the greater world, it also directly helps us in the academic sphere. It has been postulated that if more people read it, then open access will also raise the citation rate and h-index[10] [11]. In some fields, like astrophysics, the number of citations for an article roughly doubles if the articles are open access [12,13]. Although such claims are under debate in the literature, our group's experience using "open access +", particularly with citations from non-North American researchers,[11] provides some strong evidence for the effect. This effect may provide a critical advantage in the tenure process for academics providing "aggressive open access," depending on the university.

2.1.4.   Increased funding opportunities and improved student recruitment

This exposure can directly lead to funding opportunities. An example of this includes one of the research projects highlighted as part of this open-research experiment. The project looked at the effects of snow on solar photovoltaic (PV) performance. As per the dictates of the open experiment, we published our research plans on Appropedia. Due to the Appropedia posting of the details of the experiment as it was underway, the project page received the top rank in Google searches related to snow and PV technology. This exposure resulted in a successful open-source partnership being formed between 20 organizations,

---

[9]We use academia.edu although there are many other sites including repositories at many institutions themselves. For open access to all our papers whose copyright restrictions allow it, see http://mtu. academia.edu/JoshuaPearce. One interesting feature of academia.edu is that it provides access to Google Analytics for your papers, which enables you to see what search terms people use to find your articles, what sites they come from and what country they are from. It is interesting to note that despite university sites being highly ranked in Google indexing since we have been monitoring our "sources of reads", Appropedia has consistently outperformed our institutional publication page.

[10]The h-index attempts to measure both the productivity and impact of a scientist's publications. It is computed as follows: h-index is the largest number h, such that h publications have at least h citations. The h-index is often used as a tool by committees in determining tenure at many universities.

[11]Anyone who has done any collaborating with other academics at smaller universities or in the developing world knows how difficult simple access to the literature can be for some academics. Access to the complete body of up-to-date peer-reviewed literature is extremely expensive and even top research universities do not normally carry every journal in specific subfields.

FIGURE 2.2
Part of the solar photovoltaic arrays that make up the open solar outdoors test field taken from a live open-access video feed on January 9, 2013.

all of which directly contributed funding, equipment and in-kind support to produce the Open Solar Outdoors Test Field (OSOTF) (Figure 2.2), which provides critical data and research in the public domain for solar PV systems' optimization in snowy environments [14]. Unlike many other projects, the OSOTF is organized under open-source principles guided by the hacker ethic. All data and analyses when completed are made freely available to the entire PV technical community and to the general public, and already much of the data including real-time photos of the OSOTF are available in 5 min increments live.[12] The collaborating partners in the OSOTF project were willing to

---

[12]http://snowstudy.ati.sl.on.ca/.

overcome the challenges of doing open-source research because they saw value in having free access to critical research data that would be useful for product improvement, more reliable predictions of performance for funding, and reductions in solar electric system losses. It is clear from the number of collaborators who contacted the group that many of the partners would never have known about the project without the group using the open-source research approach. In addition, the coverage assisted in attracting high-quality domestic graduate student researchers.

Thus, the advantages to providing some open access to work underway for any researcher wanting to disseminate their research is clear. There are also advantages of using an open-research approach within a research group. First, the open literature reviews are extremely effective for passing on knowledge to the next generation of students, finding references when writing a paper, and keeping track of background research. These are useful to group alumni who often continue to use the wiki and contribute to the efforts of the group when they stay in the field(s). Open protocols are important for maintaining standards within the group research, but also building more easily on the work of others. In some disciplines, the sharing of free- and open-source protocols is very advanced (e.g. OpenWetWare.org, which is an effort to promote the sharing of information in biology and biological engineering), while other disciplines (e.g. mechanical engineering relying on proprietary ASME standards or electrical engineers relying on IEEE) are not. There are several generic open-source research web sites now that can provide the needs for nearly any discipline such as openresearch.org (a semantic wiki for the sciences) or myexperiment.org (which publishes scientific workflows).

There are thus advantages of greater adoption of open-access and open-source research for students. Free (zero cost) access to the latest literature is an intuitively obvious advantage, which should result in better-quality lab reports, term papers, and student writing. However, students also benefit from faster learning and higher quality information from the aggressive and potentially massive peer-review made possible by open research. Finally, there is clear evidence that combining open research with service learning in the classroom provides superior learning outcomes for students [15].

2.2. OVERCOMING CHALLENGES OF OPEN-SOURCE RESEARCH

To publish in the peer-reviewed literature, unfortunately (and perhaps for not much longer as IP in the form of patent and copyright law are being challenged [16]), authors normally must sign away IP rights (copyright). Thus, prepublishing work on the Internet before submission to a journal is a clear concern. Until IP rules are revised or completely eliminated, this risk can be avoided by limiting the type of information that is shared prior to publication. In addition, most publishers such as Elsevier (who is publishing this book) now allow the posting of preprints in institutional repositories after submission—just as the physics community does in arXiv.org.

Vandalism is a concern on open wikis. Since 2009 during the course of these experiments with open research, we saw several cases of vandalism. This was not an overly large problem, however, as they were easily and immediately corrected by Appropedia administrators as all work in a page is tracked in the history tab (Figure 2.3). Users can add pages to their watchlists (tab shown in upper middle of Figure 2.3) and can be notified if it is edited by someone else. For high levels of security, you can request administrators to "protect" pages to avoid further

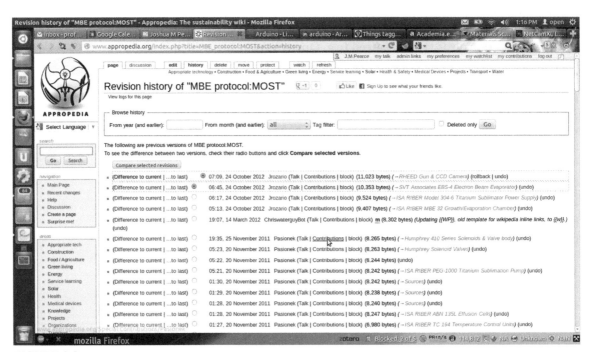

FIGURE 2.3

A screenshot of the revision history for the MOST group protocol for their molecular beam epitaxy (MBE) system used to make semiconductor layers for solar cells. Simply by clicking the "undo" next to any edit, the page can be revised, thus discouraging vandalism "softly". Users Passionek and Jrozario are student researchers in MOST, but the revision list also demonstrates how the Appropedia community assists improving the page—in this example, by using bots to automate corrections back to definitions found in Wikipedia for new users or for new students.

edits (comments are still possible on the discussion tab).[13] During the course of this experiment, these additional levels of security were found to be unnecessary. All major articles were published in the traditional peer-reviewed literature and made available via open access in institutional repositories. All the other components were left as open edit to enable future users to improve upon them.

More scientists using this approach would almost certainly improve the scale of the benefits. In a vibrant and well-populated open-source research community, maintaining literature reviews would become a community affair and not the primary work of a single group. If the majority of researchers simply added their own work, this type of initiative would be significantly improved. Similarly, in many fields, custom equipment designs, methods and software are kept private to an institution or research group—if these were provided openly in some form of web site, via a wiki protocol or "instructable", the benefits for all in the particular field would be enhanced appreciably.

The results of these open-source research experiments indicate that scientific research can be accelerated and the results disseminated faster if an open-source research methodology is used. The pragmatic, purely self-serving benefits for most academic research groups to adopt this methodology are clear. You will get more money, better students, more citations, and have a larger impact on your field. In the work that my group focuses on—encouraging sustainability via technology development—ethical considerations also play a role. Is it ethical not to provide sustainability-related research information for free to others if so doing would be more likely to create a sustainable society?

2.3. WHY SHOULD YOU SHARE AND BE NICE ANYWAY—THE THEORY

The idea that the evolutionary theory of survival of the fittest can be transferred to social interactions among humans has gained widespread acceptance in our society, most notably in some corporate cultures. Following this theory, it is argued that concepts such as charity, fairness, forgiveness and cooperation are for losers—evolutionary loose ends, soon to go extinct and of little consequence. This idea, which historically would have been considered perverse, is perhaps best summarized by the novelist Ayn Rand, when she explicitly argues that selfishness should be considered a virtue. Following this thinking, we are

---

[13]Looking carefully at Figure 2.3, you can see that my user account has administrator privileges and can thus lock pages. These extended privileges are also generally offered to other academics that use the site regularly if it is needed.

encouraged to pursue our own self-interest at all costs, to maximize not only our own outcomes but also those of all of the society. Some authors have even cynically claimed that altruism is self-interest in disguise. Anyone that has played a team sport, like basketball or football, might immediately notice a flaw in this logic based on their practical experience, yet the virtue of selfishness still holds incredible sway in our culture and has even infected many of our academic laboratories [2].

What if selfishness not only is not optimal for everyone, but it even hurts the selfish individual? New data challenge the standard model and show exactly that. For example, in *Supercooperators—Beyond The Survival of the Fittest: Why Cooperation, not Competition, is the Key to* Life, Harvard's celebrated evolutionary biologist Martin Nowak explains that cooperation is central to the 4×10^9 year old puzzle of life [17]. Consider the following scenario: You are on the sidewalk in front of your lab and notice that a small child chasing a ball is about to enter the street and get hit by an oncoming car. Your natural reaction is probably to yell and rush into the street to save the child. Good for you. But wait, if life is about survival of the fittest, why would you trouble yourself at all let alone risk your own life to run into the street to save a stranger? You do it because our ancestors had a distinct evolutionary advantage to cooperate. Selfish cavemen got left by the clans to starve on the savannah alone.[14] As Nowak shows, cooperation not competition is the defining human trait [17]. Similarly, recent work by Keltner demonstrates that humans are not hardwired to lead lives that are "nasty, brutish, and short"—we are in fact born to be good [18].

If I am working on a 3-D printer at home and drop a screwdriver, my 2-year-old son will pick it up and hand it to me. Now it is true that he is awesome and you would be right in assuming he is gifted far beyond his years; but, it turns out you can do the same experiment in front of any young child and get the same results. This is not a learned behavior as psychologist Michael Tomasello has shown through observations of experiments on young children [19]. Tomasello shows that human children are naturally—and uniquely—cooperative. Other species just do not do it. For example, when apes are run through similar experiments, they demonstrate the ability to work together and share, but choose not to. That selfishness has made them evolutionary losers—humans not apes are the dominant species on the planet. As our children grow, their hard-wired desire to help—without expectation of reward—becomes diluted and perverted by our culture. As our children become more aware of being a member of a group(s), the group's mutual expectations can

---

[14]Today selfish academics can only reply to questions at conferences "I can't answer that, it is proprietary" a few times before being left behind to perish in the libraries of ignored literature.

either encourage or discourage altruism and collaboration. *Makers*,[15] for example, have a mutual expectation of collaboration, which is why they reacted so negatively when Makerbot, a formally open-source 3-D printer company, announced that their latest printer (Replicator 2) would not come with all the source code [20] and MakerBot CEO worked so hard to maintain their reputation of "niceness" [21]. Either way, cooperation emerges as a distinctly human combination of innate and learned behaviors. The open-source philosophy thus is aligned with our natural tendency toward cooperation.

The FOSS movement we discussed in Chapter 1 has produced a community of hackers and computer programmers whose shared goal is to work together to develop better computer software [22]. Similarly, in the FOSH movement, there are burgeoning communities of hackers and makers working together to build everything from the open-source 3-D printers we will discuss in Chapter 5 and scientific research tools [23], the topic of this book to DIY drones, which is the largest amateur unmanned aerial vehicle (UAV) community in history.[16]

2.4. INDUSTRIAL STRENGTH SHARING

In the area of competition, natural selection takes place not only among individuals but also when groups compete with other groups. Here, our intuition about cooperation among teammates is correct and we have a form of group selection—the best team wins, just like in football or basketball. Even in the most cut-throat competitive business world, there is overwhelming evidence of the

---

[15]The **maker** movement is a relatively new, contemporary subculture, representing a technology-based extension of the more old-school DIY culture. Think mash up of Martha Stewart and Tony Stark (Iron Man). These modern day cyberpunks enjoy engineering-oriented pursuits for fun such as electronics, robotics, 3-D printing, and the use of CNC (computer numerical control) tools, as well as more traditional activities such as metalworking, woodworking, and traditional arts and crafts. The maker subculture stresses new and unique applications of technologies, and encourages invention prototyping, and of course sharing and building on one another's work. There is a strong focus on using and learning practical skills and applying them creatively. For more information about the Maker movement, read *Make Magazine* (http://makezine.com/) or consider visiting a local Hacker or Maker Space, which are community-operated physical places, where people can meet and work on their projects (http://hackerspaces.org/wiki/). Makers are not just playing around, although both the fictional literature (see Makers by Cory Doctorow for an excellent introduction to the subculture http://craphound.com/makers/) and reading their blogs makes it look like fun. It has already been predicted to foster the next great industrial revolution [57].

[16]See http://diydrones.com/. This community created the Arduino-based ArduPilot, the world's first universal autopilot, which can be applied to planes, multicopters of all sorts and ground rovers. We will cover much more details about the open-source Arduino electronic prototyping platform in Chapter 4. The ArduPilot Mega 2 autopilot hardware runs a variety of powerful free UAV software systems, including (1) ArduPlane, a pro-level UAV system for planes of all types and (2) ArduCopter, a fully autonomous multicopter and helicopter UAV system. The potential scientific applications of these low-cost systems have only begun to be explored.

advantage of companies cooperating with others to gain a competitive advantage over their rivals. In fact, there is a whole discipline that has grown to study these interactions: industrial ecology, which is the study of material and energy flows through industrial systems. One of the most useful concepts to emerge from this new discipline is that of industrial symbiosis. In *industrial symbiosis*, traditionally separate industries are considered collectively to gain competitive advantage by instituting the mutually beneficial physical exchange of materials, energy, water, and/or other by-products. Such a system collectively optimizes material and energy use at efficiencies beyond those achievable by any individual process alone. The key benefits of industrial symbiosis are collaboration and the synergies offered by geographic proximity [24]. Industrial symbiotic systems such as the now-classic network of companies in Kalundborg, Denmark have spontaneously evolved from a series of microinnovations of by-product sharing over a long timescale [25,26]. At the center of the Kalundborg ecoindustrial park is the 1500 MW Asnaes Power Station, which supplies surplus (1) heat to 3500 local homes/residences and a nearby fish farm, (2) gypsum from the sulfur dioxide scrubber to a wallboard manufacturer, (3) sludge for use as fertilizer on local farms, and (4) steam to a Statoil plant and Novo Nordisk, a pharmaceutical and enzyme manufacturer. This reuse of by-products (what less advanced industries call "waste") reduces landfill load and mining, and reuse of heat reduces the amount of thermal pollution discharged to a nearby fjord. Furthermore, fly ash and clinker from the power plant is used for road building and cement production. These are only a few of the dozens of corporate mutually beneficial links found in the Kalundborg system. To catalog them all is far beyond the scope of this book and unnecessary. However, it is important to understand how this type of collaboration among firms has enormous economic benefits. To illustrate this point on a smaller ecoindustrial system, consider a multi-Gigawatt solar PV factory at the center of a next generation ecoindustrial park made up of eight symbiotic factories as seen schematically in Figure 2.4 [27].

The first factory (1) is a conventional recycling facility, which is used to source the glass and aluminum needed to fabricate the solar cell from recycled materials (when viable) and thus have a lower embodied energy (95% lower for aluminum and 20% for glass) [28]. The raw glass from the recycling plant is fed to a sheet glass factory (2), which outputs cut sheets of 3-mm-thick glass with seamed edges, which provides both substrates and potentially back cladding for the PV. Finally, the glass is tempered for mechanical strength and the front side is coated with a transparent conductor such as tin oxide, zinc oxide, or indium tin oxide to be used as thin-film PV substrates [27]. The potential symbiosis of colocating the glass plant (2) and the PV plant (4) while using recycled glass is shown in Figure 2.5.

Optimization occurs if the demand for the glass materials is large enough to warrant a dedicated line that produces solar-grade glass. This is necessary

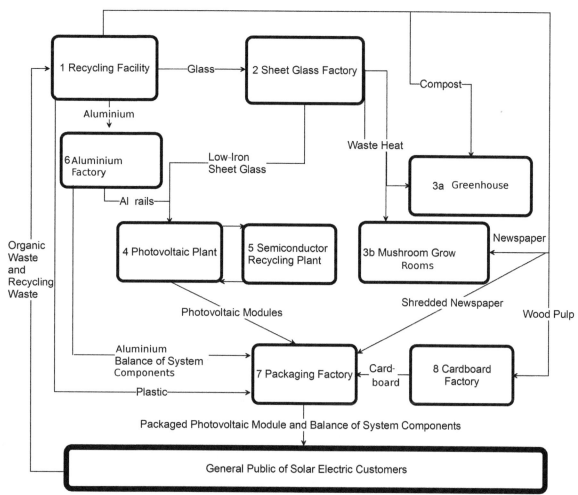

FIGURE 2.4
A schematic of an eight factory ecoindustrial park centered around a Gigawatt-scale solar photovoltaic plant. The mutually beneficial material and energy flows are shown that make up industrial symbiosis in the system.

because glass specifically manufactured with low iron content for PV cells can increase the sunlight entering the cell by about 15% and have a corresponding improvement in device performance [27]. At current solar PV manufacturing lines for thin films, altering the glass recipe for small batches is uneconomic, but this is reversed at large scales of manufacturing [29]. Historically, sheet glass was rarely customized for PV cell production, now as PV manufacturers couple with glass manufacturers, the cost of completed modules is falling rapidly [30]. The energy savings was found to be over 12%, and this symbiosis also cuts the use of 5266 ton of crude oil per year and reduces glass inputs into

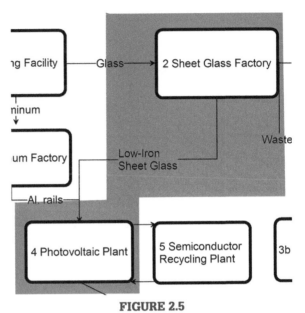

FIGURE 2.5
Details of industrial symbiosis between PV and glass manufacturing.

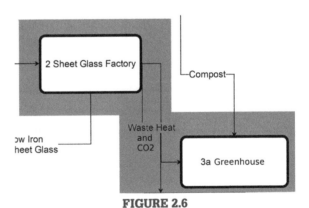

FIGURE 2.6
Industrial symbiosis between glass manufacturing and a greenhouse complex.

the factory by 30,000 ton per year [31]. These energy and materials savings can all be directly related to saving money/reducing costs.

The production stages in the glass factory that utilize large amounts of heat have integrated thermal recovery to provide lower grade heat and carbon dioxide for a multiacre "waste heat" tomato greenhouse complex (3a) shown in Figure 2.6 [32]. In the greenhouse complex, plants can be grown year round

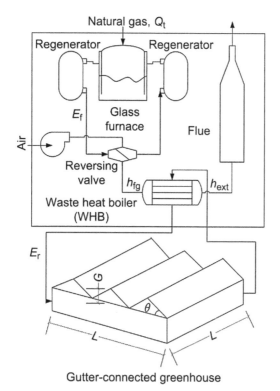

FIGURE 2.7

Details of industrial symbiosis between glass manufacturing and a greenhouse complex showing system structure and energy flows.

in even northern climates utilizing the waste heat from the manufacturing plants in the ecoindustrial park and this approach will be detailed in the second example below. Similarly, the waste heat could instead (or in addition to) be utilized to provide grow rooms for mushrooms (3b). In both agricultural plants, the food or other agricultural products is sold outside the park and the growing medium will be provided by the recycling facility (1), compost for the greenhouse (3a) and wood pulp or compost for the mushroom growing facility (3b).

The glass plant produces 500 ton of glass per day, making it a mid-sized float glass operation [33], and utilizes 1.25 PJ of natural gas a year according to the U.S. Department of Energy [34]. The entire waste heat system is shown in Figure 2.7 with the proposed structure and energy flows.

If properly treated, the carbon dioxide (CO_2) can be useful to the greenhouse complex. Modern greenhouse operations utilize CO_2 enrichment in order to increase crop yields. Particularly in a tightly sealed greenhouse being heated in the winter time, CO_2 enrichment is required at the minimum to maintain the atmospheric concentration of CO_2 at ambient levels (around 380 ppm) to account for plant photosynthesis [35]. Utilizing extremely conservative estimates, it has been found that a flat glass plant could support about a 4 acre greenhouse complex in Canada, where 700 ton of tomatoes can be grown each year and offset from 1000 to 2000 ton of CO_2 annually. Additionally it has been shown that over a 20 year campaign with a 10% MARR, the waste heat system is significantly less expensive to operate than a purely natural gas system. Finally, the addition of a waste heat greenhouse can reduce the costs of emissions compliance for a company, as the deferred costs of liquid CO_2 can fund millions of dollars of emissions reductions retrofits [32]. The bottom line of this analysis is that the waste heat system is significantly more economical to operate than a purely natural gas system from the greenhouse perspective. These economic gains can be transferred to the glass factory as these by-products (heat and CO_2) can be sold for a profit, which thus can enable further reduction in glass costs, which in turn provide the potential to reduce the costs of PV modules. These savings can provide a significant competitive advantage over PV manufacturers, which do not look beyond their own gates for synergies.

The substrates are then fed directly into the PV module plant (4), where a group of semiconductor and metal thin-film deposition systems create and pattern the active layers of the PV modules. All waste semiconductors and metals are captured and returned to a semiconductor recycling plant (5) to supplement the incoming and generally expensive high-purity materials going into the deposition systems. Using a recycling process that results in a silane loss of only 17%, instead of conventional processing that loses 85% silane, it results in an energy savings of 81,700 GJ and prevents 4400 ton of CO_2 from being released into the atmosphere per year for the single junction plant. Due to the increased use of silane for the relatively thick microcrystalline Si:H layers in the tandem junction plants, the savings are even more substantial—290,000 GJ of energy savings and 15.6 million kg of CO_2 eq. emission reductions per year. This recycling process reduces the cost of raw silane by 68%, or approximately $22.6 million per year for a 1 GW Si:H-based PV production facility and over $79 million per year for tandem manufacturing [36]. Similar arrangements are possible for the other factories shown in Figure 2.4. In each case, it is economical for the only marginally connected companies to share information and by-products.

Even within industries themselves, it can be beneficial to adopt an open-source methodology. There are literally dozens of open-source business models for software and hardware and hundreds (thousands?) of successful companies [37–50]. If we turn our attention to the PV industry used in our example, there are at least four business models that allow the industry to enjoy the benefits of open-source development [51]. Both a partnership model (where companies pull their resources to attack universal problems) and a franchise model (where resources our pooled, but competition is limited by geographic boundaries) are examples of open-source business systems that could operate within the current industry framework, while still allowing for the sharing of information and direct collaboration. Next, the secondary industry model (where a supplier of materials encourages open-source development to improve the products of their downstream businesses to grow the entire market) is a method that could be used to shift the PV industry into a more open environment. Finally, following the incredible success of the open-source software model, the entire industry could be opened, similar to Linux development, which has created billions in profit for hundreds of companies and countless jobs.

The PV industry is not some sort of anomaly. The same tactics can be applied to any industry. By colocating factories in ecoindustrial parks, both the transportation costs and transportation energy between them can be minimized and many of the inputs for the plants can literally come from waste products in the surrounding population centers. The use of open-source information to map potential energy and cost savings from transportation reductions and industrial symbiosis exists in many fields [52,53]. In addition, the wide variety of open-source business models enable essentially any form of company to take advantage of industrial-scale sharing even within the industry itself with their competitors.

2.5. THE FATE OF HARDWARE VENDORS: INNOVATE OR DIE

Often after people have been exposed to the concept of open-source hardware and understand the potential for substantial cost reductions for superior equipment, they grow concerned about the commercial enterprises that support science. For example, our group has shown how scientists can reduce the cost of optical equipment by over 97% using the open-source paradigm (details in Chapter 6.2) [54]. At first glance, this would mean economic ruin for any scientific optics company, particularly if they chose to forgo continual innovation and improvement of their products. Thus, it is tempting to think a company like Edmund Optics that has a historical large share of business in the scientific optics field would be threatened from recent advancements [54]. However, this does not appear to be true. The *Chronicle of Higher Education* reported that Edmund Optics welcomes the approach. Their view bolsters the fundamental premise of this book. Edmund Optics believes that in the near future, it is likely that many (or all) scientists and engineers will manufacture their own lab equipment using open-source hardware designs. This will undoubtedly cost the company some of its product line. However, Edmund Optics feels that successful manufacturing companies will continue to play a role in making equipment that meets higher precision standards than are possible with the current capabilities of 3-D printers [55]. So, for example, although a lens holder can be printed with a low-cost 3-D printer today, scientists would still need to buy the lens and better yet they may be willing to buy even more lenses because they can print more complicated optical experimental mechanical systems. As open-source 3-D printing continues to evolve and scientists obtain the ability to print lenses, Edmund Optics and any other scientific equipment manufacturer will be needed to continually innovate to stay in business. Edmund Optics believes they are up to the challenge. With their positive welcoming of the innovation surge brought on by the open-source hardware paradigm, it is highly probable that Edmund Optics will continue to play a leading role in scientific equipment supply. Companies that attempt to fight the tidal wave of innovation with antiquated IP-related litigation are likely to be swept away by history. The bottom line is that companies that are unable or unwilling to innovate are no longer viable.

2.6. CONCLUDING THOUGHTS

This chapter explored both the extremely pragmatic and self-serving advantages to your group for joining the open-source scientific community for your equipment and instrumentation, but it also showed how the human hard-wiring for cooperation provides both our species and our organizations and industries a distinct competitive advantage. The legal and licensing issues involved in best capitalizing on the opportunities for advanced sharing are discussed in the next chapter.

REFERENCES

[1] Lieberwitz RL. Educational law: the corporatization of academic research: whose interests are served? Akron Law Rev 2005;38(759):764–5.

[2] Chan AS, Fisher D. Exchange university: corporatization of academic culture. Vancouver: University of British Columbia Press; 2008.

[3] Gelman IJ. Opinion: missing methods. Scientist 2012; Available from: http://the-scientist.com/2012/05/03/opinion-missing-methods/ [accessed 9.05.12].

[4] Weber S. The political economy of open source software. Berkeley Roundtable on the International Economy, Working Paper 2000;140.

[5] Levy S. Hackers: heroes of the computer revolution. New York: Doubleday; 1984.

[6] Bergquist M, Ljungberg J. The power of gifts: organizing social relationships in open source communities. Open Inf Syst J 2001;11:305–20.

[7] Bessen J, Maskin E. Sequential innovation, patents, and imitation. RAND J Econ 2009;40(4):611–35.

[8] Pearce JM. Open source research in sustainability. Sustainability: J Rec 2012;5(4):238–43.

[9] Denkenberger DC, Pearce Joshua M. Compound parabolic concentrators for solar water heat pasteurization: numerical simulation. Proceedings of the 2006 international conference of solar cooking and food processing; 2006. p. 108.

[10] Denkenberger DC, Brandemuehl MJ, Pearce JM, Zhai J. Expanded microchannel heat exchanger: design, fabrication and preliminary experimental test. Proc Inst Mech Eng Part A 2012;226:532–44.

[11] Harnad S, Brody T. Comparing the impact of open access (OA) vs non-OA articles in the same journals. D-Lib Mag 2004;10(6). http://www.dlib.org/dlib/june04/harnad/06harnad.html.

[12] Metcalfe TS. The rise and citation impact of astro-ph in major journals. BAAS 2005;37:555 (astro-ph/0503519).

[13] Schwarz GJ, Kennicutt RC. Demographic and citation trends in astrophysical journal papers and preprints. BAAS 2004;36:1654.

[14] Pearce JM, Babasola A, Andrews R. Open solar photovoltaic systems optimization. Proceedings of the sixteenth annual national collegiate inventors and innovators alliance conference. 2012. p. 1–7.

[15] Pearce JM. Appropedia as a tool for service learning in sustainable development. J Educ Sustainable Dev 2009;3(1):47–55.

[16] Kinsella NS. Against intellectual property. J Libertarian Stud 2001;15(2):1–53.

[17] Nowack M, Highfield R. SuperCooperators. Canongate Books; 2011.

[18] Keltner D. Born to be good: the science of a meaningful life. W. W. Norton & Company; 2009.

[19] Tomasello M. Why we cooperate: based on the 2008 tanner lectures on human values at Stanford. MIT Press; 2009.

[20] Giseburt R. Is one of our open source heroes going closed source? [WWW Document], 2012. Available from: http://blog.makezine.com/2012/09/19/is-one-of-our-open-source-heroes-going-closed-source/.

[21] Pettis B. Fixing misinformation with information. [WWW Document], 2012. MakerBot. Available from: http://www.makerbot.com/blog/2012/09/20/fixing-misinformation-with-information/.

[22] DiBona C, Ockman S, Stone M. Open sources: voices from the open source revolution. Sebastopol CA: O'Reilly & Associates; 1999.

[23] Pearce JM. Building research equipment with free, open-source hardware. Science 2012;337(6100):1303–4.

[24] Chertow MR. Industrial symbiosis: literature and taxonomy. Ann Rev Energy Environ 2000;25:313–37.

[25] Ehrenfeld J, Gertler N. Industrial ecology in practice. The evolution of interdependence at Kalundborg. J Ind Ecol 1997;1(1):67–79.

[26] Jacobsen NB. Industrial symbiosis in Kalundborg, Denmark: a quantitative assessment of economic and environmental aspects. J of Ind Ecol 2006;10:239–55.

[27] Pearce JM. Industrial symbiosis for very large scale photovoltaic manufacturing. Renewable Energy 2008;33:1101–8.

[28] Milne G, Readon C. Embodied energy. Your home technical manual: Australia's guide to environmentally sustainable homes. Commonwealth of Australia; 2005.

[29] Keshner MS, Arya R. Study of potential cost reductions resulting from super-large-scale manufacturing of PV modules 2004. NREL/SR-520-36846.

[30] Branker K, Pathak MJM, Pearce JM. A review of solar photovoltaic levelized cost of electricity. Renewable Sustainable Energy Rev 2011;15:4470–82.

[31] Nosrat, AH., Jeswiet, J, Pearce, JM. 2009. Cleaner production via industrial symbiosis in glass and large-scale solar photovoltaic manufacturing, science and technology for humanity (TIC-STH). 2009 IEEE Toronto international Conference, p. 967–70.

[32] Andrews R, Pearce JM. Environmental and economic assessment of a greenhouse waste heat exchange. J Cleaner Prod 2011;19:1446–54.

[33] Pellegrino J, Sousa L, Levine E. Energy and environmental profile of the U.S glass industry. Office of Industrial Technologies 2002 [Final Report].

[34] U.S Department of Energy. Industrial assessment centers database. Accessed at: 2007 http://www1.eere.energy.gov/industry/bestpractices/iacs.html.

[35] Chalabi ZS, Biro A, Bailey BJ, Aikman DP, Cockshull KE. Structures and environment: optimal control strategies for carbon dioxide enrichment in greenhouse tomato crops, part 1: using pure carbon dioxide. Biosyst Eng 2002;81(No. 4):421–31.

[36] Kreiger MA, Shonnard DR, Pearce JM. Life cycle analysis of silane recycling in amorphous silicon-based solar photovoltaic manufacturing. Resour, Conserv Recycl 2013;70:44–9.

[37] Hecker F. Setting up shop: the business of open-source software. IEEE Software 1999;16:45–51.

[38] DiBona C, Ockman S, Stone M, editors. Open sources: voices from the open source revolution. O'Reilly & Associates, Inc; 1999.

[39] Raymond E. The cathedral and the bazaar. Knowl Technol Policy 1999;12:23–49.

[40] Mahadevan B. Business models for internet-based ecommerce: an anatomy. Calif Manage Rev 2000;42(4):55–69.

[41] Pomerance GM. Business models for open source hardware design 2000. Available at: http://pages.nyu.edu/~gmp216/papers/bmfosh-1.0.html.

[42] Fink M. Business and economics of Linux and open source. Prentice Hall Professional Technical Reference; 2002.

[43] Mockus A, Fielding RT, Herbsleb JD. Two case studies of open source software development: Apache and Mozilla. ACM Trans Softw Eng Methodol 2002;11:309–46.

[44] Bonaccorsi, Rossi. Why open source software can succeed. Res Policy 2003;32:1243–58.

[45] Bonaccorsi A, Rossi C, Giannangeli S. Adaptive entry strategies under dominant standards — hybrid business models in the open source software industry. SSRN J 2004 http://dx.doi.org/10.2139/ssrn.519842.

[46] Krishnamurthy S. An analysis of open source business models. In: Feller J, Fitzgerald B, Hissam S, Lakhani K, editors. Perspectives on free and open source software. Cambridge: MIT Press; 2005.

[47] Mulgan G, Steinberg T, Salem O. Wide open: open source methods and their future potential. London: Demos; 2005.

[48] Tuomi I. The future of open source. In: Wynants M, Cornelis J, editors. How open is the future? Brussels, Belgium: VUB Brussels University Press; 2005. p. 429–59.

[49] O'Reilly T. What is web 2.0: design patterns and business models for the next generation of software. Commun Strateg 2007;65(1):17–37.

[50] Zott C, Amit R, Massa L. The business model: theoretical roots, recent developments, and future research. IESE Business School University of Navarra; 2010, Working Paper-862.

[51] Buitenhuis AJ, Pearce JM. Open-source development of solar photovoltaic technology. Energy Sustainable Dev 2012;16:379–88.

[52] Pearce JM, Johnson SJ, Grant GB. 3D-mapping optimization of embodied energy of transportation. Resour, Conserv Recycl 2007;51:435–53.

[53] Doyle W, Pearce JM. Utilization of virtual globes for open source industrial symbiosis. Open Environ Sci 2009;3:88–96.

[54] Zhang C, Anzalone NC, Faria RP, Pearce JM. Open-source 3D-printable optics equipment. PLoS ONE 2013;8(3):e59840. http://dx.doi.org/10.1371/journal.pone.0059840.

[55] Basken Paul. Lab equipment made with 3-D printers could cut costs by 97%. March 29, 2013. https://chronicle.com/blogs/percolator/lab-equipment-made-with-3-d-printers-could-cut-costs-by-97.

[56] Boldrin Michele, Levine DK. Against intellectual monopoly. Cambridge: Cambridge University Press; 2008.

[57] Taylor C. Wired's Chris Anderson: today's "Maker Movement" is the new industrial revolution [TCTV]. TechCrunch; [WWW Document], 2012. Available from: http://techcrunch.com/2012/10/09/wireds-chris-anderson-todays-maker-movement-is-the-new-industrial-revolution-tctv/.

CHAPTER 3

Open Licensing—Advanced Sharing

3.1. INTRODUCTION

The intellectual property (IP) system is in many ways a reaction to the fictional tragedy of the commons [1]. The *commons* are a public space in which resources and ideas are accessible and are available to all. In the so-called tragedy of the commons, these resources are exploited to depletion as it is in everyone's best interest to use those resources as selfishly as possible. If one does not use the resources of the commons, they become uncompetitive when those resources are no longer available to them. Therefore, following the logic of selfishness, it is in everyone's best interest to snatch up as much property from the commons as they can before everyone else takes that property. If stakeholders in the commons try to work together, one of the greedier stakeholders can just circumvent the rest of the stakeholders by using up all the resources of the commons. Grazing land for cattle is the classic example of the tragedy of the commons. If there is a limited amount of grazing land for the cows and multiple cattle farmers, each farmer will attempt to maximize the land available for grazing as the farmer can then own and support more cattle. If the farmer does not attempt to maximize grazing land for his/her cattle, they will be undernourished compared to the other cattle whose farmers are maximizing their grazing land. Eventually, this drive for maximization will lead to the depletion of the grazing land leaving those farmers who had not maximized their grazing land at a disadvantage compared to the maximizing farmers. Ignoring for a moment, whether or not this is even true, it is less productive and we as a society lose in overall innovation if we are caught falling into the trap of selfishness.[1] There is an even bigger problem: the standard discourse assumes that this tragedy is also applicable to the domain of ideas.

First, it should be pointed out that this view of the anticommons has been critiqued for not understanding the complex relations between people and their

---

[1]This selfishness trap is clearly less productive overall than the sharing methodology discussed in Chapter 2.

Open-Source Lab. http://dx.doi.org/10.1016/B978-0-12-410462-4.00003-2

use of the commons. Many historical cultures managed to live for centuries without creating any form of tragedy with their common resources. You probably have good examples where you live (here we call them "parks", which are generally the most attractive locations in a city). In many of these cases, people from all over the world, representing many societies and cultures, have created social rules and institutional structures to manage common resources [2–4]. We do this to prevent the commons' tragedy and these rules do not necessarily have to come from private ownership of resources, although this has worked in some cultures. Structures and rules, such as clearly defined boundaries, collective choice agreements, and monitoring, have also helped to regulate the use of the commons in many fields. While this management has not always been perfect by any measure, it has led to the successful use of natural ecological resources without their depletion for much of our known human history.[2] The commons, which is created with the open-source methodology discussed in Chapter 2, works much in the same way. Licensing, crediting, and peer-review are examples of the social rules and institutional structures that regulate and manage the commons of open source in software, images, and academic scholarship, respectively. These rules allow for innovations to occur and even go a long way toward encouraging it, while at the same time, ensuring knowledge is not locked away in the anticommons. As progress in any of the fields of science or engineering occur, or more specifically, in science-related research equipment, it is clear that the anticommons methods of the industrial revolution are no longer the only routes to ensure ideas and products are respectfully credited and used. In fact, the antiquated anticommons methods are often counterproductive to both innovation and dissemination of good ideas, solutions, and optimized hardware.

This chapter will first examine software rights to pull in lessons from free and open-source software (FOSS) for open-source hardware (OSHW) developers. FOSS provides examples of the GNU Public License (GPL) and the rich selection of Creative Commons (CC) licenses, which are summarized and explored for application to hardware. Then, the current selection of OSHW licenses are reviewed including (1) the Tucson Amateur Packet Radio (TAPR) Open Hardware License (OHL), (2) the Chumby License, (3) the Simputer General Public License (SGPL), and (4) the CERN[3] OHL. Next, the Open Source Hardware Association's (OSHWA) principles and definition are provided as a basis for a transition to OSHW. The most important section for those new to the field is the review of best practices and etiquette for using

---

[2]Contrast that to distinctly anticommons approaches that are devastating the environment today. It is also perhaps useful to note what happens to civilizations that pillage their environments for short-term gain. For a detailed and sobering account of the results of that path, see Diamond's *Collapse* [38].
[3]The name CERN is derived from the acronym for the French "ConseilEuropéen pour la Recherche Nucléaire". In English it is known as the European Council for Nuclear Research.

OSHW. Finally, the chapter concludes with a review of the current IP challenges to those wishing to pursue OSHW development for the benefit of everyone and concluding thoughts are provided.

3.2.   LEARNING FROM SOFTWARE: SOFTWARE RIGHTS

Although software was born free, it was quickly locked down by IP laws. In the computer software field, where it is trivially easy to copy code and then disseminate programs for zero cost, the old industrial model of IP has been used for rent seeking[4]. To enforce copyright restrictions and limit their liability, early software developers and companies created custom licenses that were contracts between the user and developer. These expensive contracts, as they were often one-of-a-kind, were typically created by the legal representatives of both parties for each piece of software. When there were only a few companies using computers, this was cumbersome but manageable. However, as software products have become more ubiquitous and the cost of these contracts went from absurd to prohibitive, developers adopted the *End User License Agreement* (*EULA*) as the primary form of licensing. You are probably only vaguely aware of EULAs. These are the documents, and increasingly screens or hyperlinks, you agree to look at in order to use commercial software. These documents are generally written in dense legal speak, often making use of font-based tactics to make them difficult to read or understand. For example, they often use small font or all caps. When combined, this is a particularly effective method to slow readers down to increase the probability that they will give up and simply ignore what you write. Not surprisingly, the vast majority of commercial software users simply ignore EULAs and remorselessly lie when they click the button indicating they have "read and agree" to it. EULAs treat the software as a copyrighted product to which the user only has certain rights even if they have just purchased the software. Software companies use EULAs because unlike custom licenses, EULAs can be applied to every piece of software without much modification, allowing for the high-volume distribution of software. The most common EULA is the so-called "shrink-wrap agreement", in which the user of the software agrees to the license simply by opening the software package or using it [5]. While some have critiqued shrink-wrap agreements for being unenforceable [6], or for unfairly trapping users into unread agreements [7], this licensing in the world of proprietary software is ubiquitous.

---

[4]For the most part despised by economists, rent seeking is an attempt to profit by obtaining economic rent by manipulating the social or political environment of the economy rather than creating new real wealth.

Licenses are still just as important in the world of open-source software.[5] However, contrary to proprietary software, early open-source software did not include any licenses because there was no need to enforce copyright restrictions. Released into the public domain, software was free to be redistributed, used and modified as users wished. Ideally, this system could have stayed in place and everyone simply could have concentrated on the technical aspects of making superior code. Unfortunately, this anarchic system was born in a culture of greed [8], which enabled some users to repackage modified open-source software as proprietary software. Thus, software that had once been in the commons was locked back down and became private property. You can imagine how the people that wrote the original code felt. At the same time, the lack of open-source contracts encouraged companies like IBM, who worked on commons software projects such as UNIX, to assert intellectual monopoly "rights" on commons software [9]. To prevent this movement of software from the commons, simply to enrich rent seekers, a new type of licensing was necessary. Luckily, Richard Stallman, an MIT programmer and powerful supporter of the open-source movement, called for developers to release their software under the GPL[6] rather than into the public domain [10, 11].

The GPL[7] requires that the source code of a software program is made freely available to and modifiable by everyone. The GPL contains two critical vital components of the license: first, it ensures that programs released under the GPL cannot become proprietary, and in addition, the GPL prevents the commercialization of software by requiring that derivative works be subject to the same license conditions as the original program, and that open and closed software could not be mixed under the license [12]. While the GNU license was widely used in the 1980s, alternatives to the GPL were created in the 1990s to overcome some of its restrictions that some companies viewed as a problem, like the mixing of open and closed code [9]. The GPL remains the standard for true freedom and open-source development. Today, there is a wide variety of open-source software licenses that fall into three major categories: unrestrictive, restrictive, and highly restrictive [10]. Highly restrictive licenses (the good kind of restriction), like the GPL, do not allow for any mixing of open and

---

[5]The entire concept of IP may appear so absurd to some readers so as not to be worth the bother as it will likely be gone in the future. This sentiment, often expressed by more energetic graduate students, was summarized in a tweet by James Governor, founder of analyst firm RedMonk, "younger devs today are about POSS—post open-source software. f*** the license and governance, just commit to github"— https://twitter.com/monkchips/status/247584170967175169. We are, however, at best in some form of transitional state now, and to move technological governance in the right direction, licenses are critical. If innovators are not careful, their innovations will simply be locked down in the old models.
[6]GNU itself is a word hack—a recursive acronym that stands for "GNU's Not Unix".
[7]http://www.gnu.org/licenses/.

Table 3.1 Comparison of Open-Source License Classes

	Unrestrictive (e.g. BSD, X11, MIT, Python)	Restrictive (e.g.: MPL, LGPL)	Highly Restrictive (e.g. GPL)
Can be relicensed?	No	No	No
Can distribute derived works without disclosing modifications?	Yes	No	No
Can incorporate in a combined work with closed source files?	Yes	Yes	No

Source: Adapted from Bruns [13].

closed code or any transfer of open code to closed code. Unrestrictive licenses (e.g. Berkeley Software Distribution (BSD)) are more tolerant of losing the "freedom" of the code. Table 3.1 [13] shows the major differences between the three classes of open-source software licenses. In the case of unrestrictive licenses, the line between open and closed software licenses is very close, but distinct.

Today, the open-source definition has matured and open-source software licenses must meet 10 fundamental requirements [14]:

1. The license must not restrict commercial or noncommercial (NC) distribution of software in any way.
2. The license must require that the software be distributed with compiled and clear source code.
3. The license must allow modifications and derived work to be distributed under the same terms as the original work.
4. The license must maintain the integrity of the author's source code.
5. The license must not discriminate against any person or group of persons.
6. The license must not restrict anyone from making use of the program in a specific field of endeavor.
7. The license must apply to everyone receiving the redistributed program without the need for an additional license.
8. The license must not be specific to a product.
9. The license must not restrict software distributed under other licenses but in the same package.
10. The license must be technology neutral.

For details of each, see the *Open Source Definition*.[8] All these requirements must be met for a license to be considered a true open-source license, not just some of them. A list of open-source software licenses is maintained by the Open

---

[8]http://opensource.org/docs/osd.

Source Initiative.[9] However, open-source licenses are not limited to software. The concept of the open-source license has spread to other areas like documentation,[10] art,[11] and academic journals.[12] These licenses have the same major characteristics as open-source software licenses: the availability of the "source code" and the right to free redistribution.

3.3.   OSHW LICENSES

Some of these licenses could obviously be easily adapted to hardware.

3.3.1.   TAPR Open Hardware License

An example of this is comes from the amateur radio community. The TAPR organization is an international amateur radio organization that has developed the TAPR OHL.[13] The TAPR OHL provides a framework for licenses for physical artifacts that is synonymous with the more established treatment of open-source software [15]. Unfortunately, licensing for hardware is complicated by the fact that the established legal concepts that are well suited for software (such as copyright and copyleft) are not as easily applied to hardware products and the documentation used to create them, although as we will see below this approach is used on Thingiverse[14], an open repository of digital designs.

The TAPR OHL preamble explains how the license works and how to use it [15]:

> "Open Hardware is a thing - a physical artifact, either electrical or mechanical - whose design information is available to, and usable by, the public in a way that allows anyone to make, modify, distribute, and use that thing. In this preface, design information is called "documentation" and things created from it are called "products."

> The TAPR Open Hardware License ("OHL") agreement provides a legal framework for Open Hardware projects. It may be used for any kind of product, be it a hammer or a computer motherboard, and is TAPR's contribution to the community; anyone may use the OHL for their Open Hardware project. You are free to copy and use this document provided only that you do not change it."

---

[9]http://www.opensource.org/licenses/alphabetical.
[10]GNU Free Documentation License http://www.gnu.org/copyleft/fdl.html.
[11]Open-art license http://three.org/openart/license/.
[12]Open-access journals http://www.doaj.org/.
[13]TAPR open-hardware license http://www.tapr.org/ohl.html.
[14]Thingiverse http://www.thingiverse.com/.

Similar to the GPL, the OHL is designed to guarantee your freedom to share and to create. It forbids anyone who receives rights under the OHL to deny any other licensee those same rights to copy, modify, and distribute documentation, and to make, use and distribute products based on that documentation. The OHL is not a complete analogy to the GPL. Unlike the GPL, the OHL is not primarily a copyright license. While copyright protects documentation from unauthorized copying, modification, and distribution, it has little to do with your right to make, distribute, or use a product based on that documentation. For better or worse,[15] patents play a significant role in those activities. Although it does not prohibit anyone from patenting inventions embodied in an open hardware design, and of course cannot prevent a third party from enforcing their patent rights, those who benefit from an OHL design may not bring lawsuits claiming that design infringes their patents or other IP. The OHL addresses unique issues involved in the creation of tangible, physical objects and hardware, but does not cover software, firmware, or code loaded into programmable devices. Thus, unnecessary complexity ensues as a copyright/copyleft-oriented license such as the GPL is better suited for these types of innovations. The reader will unquestionably see the looming legal mess of obtaining the appropriate licenses for a complex scientific tool made up of hardware, firmware and software.

In general, however, you can use the OHL (or an OHL design) if the following apply:

- You may modify the documentation and make products based upon it.
- You may use products for any legal purpose without limitation.
- You may distribute unmodified documentation, but you must include the complete package as you received it.
- You may distribute products you make to third parties if you either include the documentation on which the product is based or make it available without charge for at least 3 years to anyone who requests it.
- You may distribute modified documentation or products based on it, if you
  - license your modifications under the OHL;
  - include those modifications, following the requirements stated below;
  - attempt to send the modified documentation by email to any of the developers who have provided their email address. This is a good faith obligation—if the email fails, you need do nothing more and may go on with your distribution.

---

[15]For a detailed treatment of the "for worse," see *Against Intellectual Monopoly* by Michele Boldrin and David Levine [23]. Even the idea of patents has been being challenged by ever increasingly formidable mountain of evidence and Boldrin and Levine offer a good introduction to the literature.

- If you create a design that you want to license under the OHL, you should do the following:
  - Include the OHL document in a file named LICENSE.TXT (or LICENSE.PDF) that is included in the documentation package.
  - If the file format allows, include a notice like "Licensed under the TAPR OHL (www.tapr.org/OHL)" in each documentation file. While not required, you should also include this notice on printed circuit board artwork and the product itself; if space is limited, the notice can be shortened or abbreviated.
  - Include a copyright notice in each file and on printed circuit board artwork.
  - If you wish to be notified of modifications that others may make, include your email address in a file named "CONTRIB.TXT" or something similar.
- Any time the OHL requires you to make documentation available to others, you must include all the materials you received from the upstream licensors. In addition, if you have modified the documentation, you should do the following:
  - You must identify the modifications in a text file (preferably named "CHANGES.TXT") that you include with the documentation. That file must also include a statement like "These modifications are licensed under the TAPR OHL."
  - You must include any new files you created, including any manufacturing files (such as Gerber files) you create in the course of making products.
  - You must include both "before" and "after" versions of all files you modified.
  - You may include files in proprietary formats, but you must also include open format versions (such as Gerber, ASCII, Postscript, or PDF) if your tools can create them.

For a full treatment and text of the TAPR OHL, see http://www.tapr.org/ohl.html.

3.3.2. Chumby license

Another OSHW license was developed by Chumby HDK (2013) [16]. Simply stated:

> Under this Agreement, as set out below, chumby grants you a license to use the chumby HDK to hack your chumby Device. In return, we ask that you: keep the chumby Service on an even playing field with any other service you want to point your chumby Device to; grant us a license related to your modifications and derivatives, *when and if* you make them available to others; and agree to the other terms below.

3.3.3. Using open-source software licenses for open hardware

The last two subsections covered two good examples of OSHW licenses, neither are quite perfect, so what many open-source designers have done is simply substituting in an open-source software license such as the GPL, CC, MIT, BSD and similar open licenses. These licenses are reasonable for firmware or CAD drawings and thus Thingiverse (which has over 95,000 designs in May 2013 and is growing rapidly) uses the following licenses: CC licenses of the CC-BY, CC-BY-SA, CC-BY-ND, CC-BY-NC, then BY-NC-SA a BY-NC-ND, CC-public domain dedication, GNU-GPL, LGPL and BSD license.

The CC licenses are summarized in Figure 3.1. There is obviously a lot of disagreement about licenses, which are highly varied depending on people's personal preferences. To be considered "truly open-source" and to fit in with the established definition, there need to be no restrictions on the use of the design, which includes NC, or such phrases such as "commercialization requires purchasing a license," "no government/military use," etc. The NC license, in particular, is perhaps easy to see the appeal for designers to use (e.g. people do not mind sharing, but they also feel cheated if you take what they share and then go and make a lot of money off of it). Licenses such as the CC-BY, CC-BY-SA, GPL, BSD, and MIT do not have these restrictions and are thus the "purest" form of open source.

It is easy to be sympathetic to people that do not want the creation they give to the world to be used by someone else to make a lot of money—or worse to be used to kill people. From an academic researcher's perspective, there are a couple of points you need to keep in the back of your mind when considering licenses. First, your chances of making any serious money on an innovation are extremely small unless you devote an enormous amount of time into the business end of it. Most scientists and engineers find this somewhat boring and tedious. Working on a business prevents them from doing as much innovative work on the technical side, which is why most academics avoid it. There is an unstated assumption that academics have given up crass economic compensation for academic freedom and higher pursuits in the ivory tower.[16] This is why the vast majority of our work is essentially open sourced in the literature.

Some academics do start spin-off companies from their research. Does this preclude them from participating in the open-source community? Absolutely not. Even if you are considering starting a company, the open-source methodology can be one of your greatest assets. It is often far superior to harvest

---

[16]No one would argue that most academic scientists and engineers are destitute, but there is normally more money, but less freedom on the corporate side of the fence.

The Licenses

Attribution
CC BY

This license lets others distribute, remix, tweak, and build upon your work, even commercially, as long as they credit you for the original creation. This is the most accommodating of licenses offered. Recommended for maximum dissemination and use of licensed materials.

Attribution-ShareAlike
CC BY-SA

This license lets others remix, tweak, and build upon your work even for commercial purposes, as long as they credit you and license their new creations under the identical terms. This license is often compared to "copyleft" free and open source software licenses. All new works based on yours will carry the same license, so any derivatives will also allow commercial use. This is the license used by Wikipedia, and is recommended for materials that would benefit from incorporating content from Wikipedia and similarly licensed projects.

Attribution-NoDerivs
CC BY-ND

This license allows for redistribution, commercial and non-commercial, as long as it is passed along unchanged and in whole, with credit to you.

Attribution-NonCommercial
CC BY-NC

This license lets others remix, tweak, and build upon your work non-commercially, and although their new works must also acknowledge you and be non-commercial, they don't have to license their derivative works on the same terms.

Attribution-NonCommercial-ShareAlike
CC BY-NC-SA

This license lets others remix, tweak, and build upon your work non-commercially, as long as they credit you and license their new creations under the identical terms.

Attribution-NonCommercial-NoDerivs
CC BY-NC-ND

This license is the most restrictive of our six main licenses, only allowing others to download your works and share them with others as long as they credit you, but they can't change them in any way or use them commercially.

FIGURE 3.1

Summaries of creative commons licenses. *Source:* http://creativecommons.org/licenses/.

the benefits of moving toward open source as discussed in the previous chapter. In many cases, a small start-up company does not have the financial scale to wage and win a battle in patent court. Consider the views of Nathan Seidle, the CEO of Sparkfun (a large open-source electronics company). Seidle has pointed out in discussion what he calls "IP obesity": "If your idea is unique, easily copied, and can be sold for profit in a local market, it will be." [17]. At the time of this writing, Sparkfun, an OSHW company, had 135 employees, $75 million of sales, 600,000 customers and had open sourced 430 unpatented products. Clearly, it is possible to base a business off of the open-source paradigm, but it demands constant innovation to stay ahead of the copiers. As Seidle points out, companies that do not innovate, particularly in technology, will just be overrun, with or without patents [17]. Most researchers are not in equipment design for a start-up company, and thus it is tempting to join the "beyond licenses" crowd, which simply shares their code or designs with no thought to license. This is far more common than you might think. The OSHWA conducted a survey in February of 2012 of the international OSHW community; out of more than 2000 responses from 70 countries, it was clear that the majority of the community releases designs with no license whatsoever [18]. This may prevent having to think about the issue, but what it risks is that your work could be locked away in the intellectual anticommons and slow development for others. If no licensing scheme is set up, OSHW could move to the anticommons over time. As Catarina Mota, the co-founder of Everywhere Tech[17] (open-source technology transfer), openMaterials[18] (open-source and DIY experimentation with smart materials) and altLab[19] (Lisbon's hackerspace) have pointed out, IP is a hot topic and quickly evolving. To overcome these challenges, Mota and others of the OSHWA are attempting to clarify the IP space for innovators interested in the open-source methodology.

3.3.4. Other open-source hardware licenses

There are two other main historical OHLs. First, the SGPL[20] was developed for the Simputer, which is a low-cost portable alternative to PCs. The simputer project was meant primarily to target the digital divide and help those in the developing world, as it ensures that illiteracy is not a barrier to handling a computer and enable the benefits of modern IT to reach the common man. As has been discussed elsewhere, this access can be extremely beneficial [19–21]. On the other end of the target user, technical sophistication spectrum is the CERN[21] OHL. The CERN OHL[22] governs the use, copying, modification and distribution of hardware design documentation, and the manufacture and distribution of products.

3.4. OPEN SOURCE HARDWARE ASSOCIATION DEFINITION

The OSHWA board along with the help of OSHWA members developed the OSHW definition on, appropriately enough, a wiki (freedomdefined.org).[23] The updated text of the definition can be found on the OSHWA web site and the following two sections detail the OSHWA version 1.0 of an OSHW definition [22]. The following two subsections relay both the OSHWA statement of principles and their definition in full.

---

[17]Everywhere Tech http://everywheretech.org/.
[18]OpenMaterials http://openmaterials.org/.
[19]AltLab http://altlab.org/.
[20]Simputer license http://www.simputer.org/simputer/license/.
[21]CERN is the European Organization for Nuclear Research, which employs thousands of scientists directly and hosts many more visiting scientists every year.
[22]CERN OHL http://www.ohwr.org/projects/cernohl/wiki.
[23]Open-source hardware (OSHW) http://freedomdefined.org/OSHW.

3.4.1.   Open-source hardware statement of principles 1.0

Open source hardware is hardware whose design is made publicly available so that anyone can study, modify, distribute, make, and sell the design or hardware based on that design. The hardware's source, the design from which it is made, is available in the preferred format for making modifications to it. Ideally, open source hardware uses readily-available components and materials, standard processes, open infrastructure, unrestricted content, and open-source design tools to maximize the ability of individuals to make and use hardware. Open source hardware gives people the freedom to control their technology while sharing knowledge and encouraging commerce through the open exchange of designs [22].

3.4.2.   Open-source hardware definition 1.0

The OSHW definition 1.0 is based on the open-source definition for open-source software. That definition was created by Bruce Perens and the Debian developers as the Debian Free Software Guidelines.

Introduction

OSHW is a term for tangible artifacts—machines, devices, or other physical things—whose design has been released to the public in such a way that anyone can make, modify, distribute, and use those things. This definition is intended to help provide guidelines for the development and evaluation of licenses for OSHW.

Hardware is different from software in that physical resources must always be committed for the creation of physical goods. Accordingly, persons or companies producing items ("products") under an OSHW license have an obligation to make it clear that such products are not manufactured, sold, warranted, or otherwise sanctioned by the original designer and also not to make use of any trademarks owned by the original designer.

The distribution terms of OSHW must comply with the following criteria:

1.   Documentation

The hardware must be released with documentation including design files, and must allow modification and distribution of the design files. Where documentation is not furnished with the physical product, there must be a well-publicized means of obtaining this documentation for no more than a reasonable reproduction cost, preferably downloading via the Internet without charge. The documentation must include design files in the preferred format for making changes, for example, the native file format of a CAD program. Deliberately obfuscated design files are not allowed. Intermediate forms analogous to compiled computer code—such as printer-ready copper artwork from a CAD program—are not allowed as substitutes. The license may require that the design files are provided in fully documented open format(s).

2. Scope

The documentation for the hardware must clearly specify what portion of the design, if not all, is being released under the license.

3. Necessary software

If the licensed design requires software, embedded or otherwise, to operate properly and fulfill its essential functions, then the license may require that one of the following conditions are met: (1) The interfaces are sufficiently documented such that it could reasonably be considered straightforward to write open-source software that allows the device to operate properly and fulfill its essential functions. For example, this may include the use of detailed signal timing diagrams or pseudocode to clearly illustrate the interface in operation. (2) The necessary software is released under an OSI-approved open-source license.

4. Derived works

The license shall allow modifications and derived works, and shall allow them to be distributed under the same terms as the license of the original work. The license shall allow for the manufacture, sale, distribution, and use of products created from the design files, the design files themselves, and derivatives thereof.

5. Free redistribution

The license shall not restrict any party from selling or giving away the project documentation. The license shall not require a royalty or other fee for such sale. The license shall not require any royalty or fee related to the sale of derived works.

6. Attribution

The license may require derived documents, and copyright notices associated with devices, to provide attribution to the licensors when distributing design files, manufactured products, and/or derivatives thereof. The license may require that this information be accessible to the end user using the device normally, but shall not specify a specific format of display. The license may require derived works to carry a different name or version number from the original design.

7. No discrimination against persons or groups

The license must not discriminate against any person or group of persons.

8. No discrimination against fields of endeavor

The license must not restrict anyone from making use of the work (including manufactured hardware) in a specific field of endeavor. For example, it must not restrict the hardware from being used in a business, or from being used in nuclear research.

9. Distribution of license

The rights granted by the license must apply to all to whom the work is redistributed without the need for execution of an additional license by those parties.

10. License must not be specific to a product

The rights granted by the license must not depend on the licensed work being part of a particular product. If a portion is extracted from a work and used or distributed within the terms of the license, all parties to whom that work is redistributed should have the same rights as those that are granted for the original work.

11. License must not restrict other hardware or software

The license must not place restrictions on other items that are aggregated with the licensed work but not derivative of it. For example, the license must not insist that all other hardware sold with the licensed item be open source, nor that only open-source software be used external to the device.

12. License must be technology neutral

No provision of the license may be predicated on any individual technology, specific part or component, material, or style of interface or use thereof.

3.5. BEST PRACTICES AND ETIQUETTE FOR USING OSHW

Best practices for OSHW can perhaps best be summarized by the Golden Rule (or the ethic of reciprocity): Do unto others as you would have them do unto you.[24] You want to make everything available for others that will enable them to copy and build on your work as easily as possible.

---

[24]Some version of the Golden Rule can be found in almost every ethical tradition [39], but many will be most familiar with the phrasing from Jesus (Matthew 7:12 or Luke 6:31).

For hardware projects, you need to ensure that you are including your original design files, preferably created with FOSS in an open format. This is so others can make modifications of your work without having to go through unnecessary and time-consuming reverse engineering. Make your design files as easy to understand as possible, comment/document as much as possible, organize in some sort of logical sequence and note any departures from standard manufacturing procedures. In addition to the raw files, provide any other files that would be useful for others such as STL files for use in 3-D printers of designs. Although the design files may provide all the materials and subcomponents, it is generally a good idea to include a bill of materials (BOM). Again, ideally the BOM should not only include the names of parts and part count but also the vendors/suppliers with prices and urls. The BOM can be a spreadsheet (e.g. ODS[25]) or be maintained in a wiki. In general, make your hardware project as easy to replicate anywhere throughout the entire world by choosing standard, widely available parts in your design whenever possible, and when you are not using a tool like a 3-D printer, which enables others to make custom parts with ease. Obviously, make any firmware or software available following the open-source principles and document/comment on it as well. For hardware projects, you should also include pictures of the design and document with pictures at appropriate times during its assembly or fabrication. This book provides (hopefully clear) documentation to be used as an example for some of the OSHW builds, such as the environmental chamber, in the next chapter. It is best to house instructions on an open wiki, so others may add to it and improve them. It is also helpful to link to datasheets for the components and parts of your hardware and to list the tools required to assemble it. For projects requiring specialized tools or equipment, also list sources and prices. Finally, share your designs in an appropriate centralized and easily accessible online location such as GitHub[26], Thingiverse[27], Appropedia[28], and so on, in addition to housing them on your personal web site. All files (design, BOM, assembly instructions, code, movies of installation, etc.) should be version controlled whenever possible and under an appropriate open license.

---

[25]ODS, OpenDocument Spreadsheet file format. This is used by the free and open-source Open Office (http://www.openoffice.org/) and Libre Office (http://www.libreoffice.org/). Whenever possible, use open formats so that anyone can use FOSS to gain access.

[26]Github https://github.com/. An excellent example of how to use Github for this type of work is underway by Pioneer Valley Open Science, which promotes and hosts science hackathons, workshops, and collaborations in the Pioneer Valley area of Western Massachusetts in addition to posting OSH on Github at https://github.com/Pioneer-Valley-Open-Science/pioneer-valley-open-science.github.com/wiki.

[27]Thingiverse http://www.thingiverse.com/. Many research groups maintain collections on Thingiverse of laboratory equipment. For the collection I maintain with most of the examples that Chapter 6 uses, see http://www.thingiverse.com/jpearce/collections/open-source-scientific-tools.

[28]Appropedia http://www.appropedia.org/. For an example of how to use Appropedia to maintain an OSH hub, see the MOST group method page at http://www.appropedia.org/Category:MOST_methods.

When using OSHW designs, some basic etiquette is necessary as well. Even if you are not legally bound to do so, respect the wishes of the makers that came before you as determined by the license that they chose. For example, CC-BY-SA demands that you share your derivatives back with the community. Similarly, many makers will share the plans to build a device with the world while also making it for sale. These entrepreneurs normally request that you do not go into business against them simply making copies of their product. You are, of course, free to make it for yourself and your lab, but if you want to create a business off of the idea, make sure that you have made a substantial contribution to making the product better. Always provide attribution and thanks for the work of others that you use. It costs you nothing to thank people in the acknowledgments of your papers, and it is a good way to encourage additional open-source design work for scientific equipment if we all make it a habit of thanking everyone that helped your research succeed. It is also a good idea to send a quick thank you email to designers whose work you appreciate and, if it is appropriate, give back to them through voluntary donations to continue development.

3.6. CONTINUED IP CHALLENGES

Current IP laws, which were ostensibly set up to encourage innovation, are coming increasingly under harsh scrutiny. An overwhelming barrage of carefully executed studies and articles have found weak or no evidence that IP increases innovation and, in fact, IP law often hinders it [23]. It is hard to imagine a truly advanced civilization arbitrarily holding back individuals to make use of the best possible combination of technologies, because someone else thought of a similar "embodiment" earlier. We find it difficult to imagine, because it is obvious that it is simply inefficient not to use the best technology available at the time to solve a problem. The following other inefficiencies elucidate the overall conclusion that IP (both copyright and patents) are actively hampering innovation:

1. Higher transaction costs due to IP for information exchange slows technical progress [24].
2. The inconsistently flexible "nonobvious" requirement of patents (which at this point is essentially the subjective opinion of specific patent examiners) locks away common-sense approaches in the anticommons to solving problems, and basic, obvious algorithms for creating innovations [25].
3. Patenting of building block technologies holds back downstream research and development, potentially crippling entire fields for 20-year stretches [26].
4. Many patents are simply not used, but only prevent others from following lines of inquiry [27].

Considering the current rather sobering (perhaps even terrifying) state of world affairs[29], where there are viable technical solutions to many of our global challenges, it is a tragedy to allow these inefficiencies to retard innovation. This "IP tragedy" is acute in many fields, like that of nanotechnology, where an open-source model has also been suggested to put innovation back on the fast track [13,28–34]. For now, however, IP law still exists and provides numerous problems for the open-source community.

The largest problem with the existence of IP laws and the open-source community's attempts to maintain a fertile public domain or commons is the uncertainty of the solution. The Achilles' heal of the available open-source licenses, which were discussed in the last several sections, is that they are largely new and untested in court. There remain many questions as to whether the licenses can be legally enforced [35]. There is also the added complexity that as open-source "products" are often created with the combination of ideas from many people (perhaps even under multiple inconsistent licenses), there would be difficulties in determining legally recognized contributions. Making matters even more convoluted, there are also unanswered questions from the contract law and property rights domains, as open licenses do not adhere strictly to either type of legal domain. McGowan points out that those currently open licenses would have to give way to the Copyright Act and its IP rights in a court of law [35]. Copyright law in the United States, perhaps more than other domains of IP, has been completely overrun by special interests and their lobbyists to the detriment of the public domain. Without further comment on the absurdity of it all, I submit as evidence the following fact and leave it to the reader to decide for themselves whose interests are best served by such laws: *copyright in the United States lasts 70 years after the death of author.* On the other hand, if a work has been authored by an immortal corporation, then the copyright lasts 95 years from publication or 120 years from creation (whichever expires first).

For most of OSHW design (excluding documentation and copyright of decorative elements), patent law is the primary concern. Unlike copyright, which is automatic, patents actually involve an investment (usually quite substantial[30]) to acquire. In order for the Patent and Trademark Office to allow a patent, the invention must be new, useful, and nonobvious.[31] The moral contract for

---

[29]See any of the recent publications in *Science, Nature* or the popular press detailing the risks of climate destabilization, species loss/extinction, drug resistant diseases, poverty, financial crisis/risk of global economic collapse, proliferation of weapons of mass destruction, and resource scarcities (e.g. water) among many other large-scale global problems.

[30]Normally, a patent through lawyers costs somewhere around $20,000.

[31]In the author's opinion, the last requirement is particularly weak as patents are built off of prior ideas and nature.

obtaining a 20-year monopoly on the idea is that the inventor fully discloses information that would allow others to utilize the invention. The practice today is that patents are legal documents with very limited technical use.[32] They are written by lawyers for other lawyers to read, and their purpose is to maximize legal scope, while remaining seemingly limited enough to obtain the monopoly rights.

This is a problem for the innovators—the scientists and engineers—because patents themselves have become an intellectual minefield. There is no exception for people to independently make the patented object in patent law. There is no fair use, no exception for home use or for copying objects for personal use. Worse yet, most complicated products (e.g. our favorite hardware for scientific experiments) is not covered by one patent, but probably dozens (both from combination patents that put already patented ideas together in a new way and by the classic patent thicket[33] where many patents actually have overlapping claims). Once someone has acquired a patent, all copies, *regardless of the copier's knowledge of the patent*, infringe upon that patent. So if you make anything that has been patented, you are infringing and could, in theory, be sued. It gets worse, as even using a patented device without authorization infringes on the patent. In addition, the line between repair and reproduction is murky, and may become an area of increased scrutiny as the use of 3-D printing (discussed in Chapter 5) to replace parts expands [36]. If you work at a university, a technology transfer officer or university lawyer might help you wade through the quagmire of patents and complete any paperwork for you, but resources spent on legal requirements are not being invested in your lab on actual innovation. Most innovators ignore the mess as, as of now, corporations are not coming after individual patent infringers who do not attempt to enter the market with products, and any time or money invested in paying attention to these patent issues is lost to innovation. This entire line of inquiry becomes more important as 3-D printing and digital manufacturing become common place and challenge the current IP paradigm. If you are interested in this area, I recommend Weinberg's introductory article *It Will Be Awesome If They Don't Screw It Up* [36]. Reassuringly, and perhaps surprisingly, it is concluded that—within the UK at least—private 3-D printer owners making items for personal use and not for gain are exempt from the vast majority of IP constraints, and that commercial users, though more restricted, are less so that might be imagined [37].

---

[32]Readers will, for example, find even the most poorly written instructables substantially more useful for making an object than a standard patent description of the vast majority of modern "inventions".

[33]A patent thicket is a pejorative term used to describe a dense web of overlapping patent rights that demands you reach licensing deals for multiple patents from multiple sources in order to develop a technology or bring it to market.

3.7. SUMMARY AND CONCLUSIONS

IP law is quickly evolving and mutating (or perhaps even on its way to extinction?) as both the challenges to the entire concept expand at the same time new technologies (such as 3-D printing, which we will discuss in Chapter 5) make the policing of IP-related transgressions, if not impossible, improbable and highly intrusive. There is still a great need for bolstering the legal position of the public domain and creating defensive publications to protect the intellectual commons from the last of the poachers. The legal protections to do this provided by the currently practiced IP licenses for OSHW are unclear. The OSHWA is working to develop a simple method (following CC) for innovators to understand, select and apply OSHW licenses to accelerate innovation diffusion and drive up the adoption. As of this writing, they are not ready and although the licenses we have now have unproven legal validity, they act as a social mechanism, which can assist innovators as it provides entrance into the open-source community (the numerous benefits of which were outlined in Chapter 2 and the many examples in the next several chapters). If you follow the Golden Rule and respect the rights of other innovators, the open-source community will repay any investment you make many times over.

REFERENCES

[1] Hardin G. The tragedy of the commons. Science 1968;162(3859):1243–8.

[2] Ostrom E. Governing the commons: the evolution of institutions for collective action. Cambridge University Press; 1990.

[3] Ostrom E, Walker J. Communication in a commons: cooperation without external enforcement. Laboratory research in political economy. 1991. p. 287–322.

[4] Dietz T, Ostrom E, Stern PC. The struggle to govern the commons. Science 2003;302(5652): 1907–12.

[5] Gomulkiewicz RW, Williamson ML. A brief defense of mass market software license agreements. Rutgers Comput Technol Law J 1996;22(1):335–68.

[6] Lemley MA. Intellectual property and shrinkwrap licenses. South Calif Law Rev 1994;68(1): 1239–94.

[7] Goodman B. Honey, I shrink-wrapped the consumer: the shrink-wrap agreement as an adhesion contract. Cardozo Law Rev 1999;21:319.

[8] Sims RR, Brinkmann J. Enron ethics (or: culture matters more than codes). J Bus Ethics 2003;45(3):243–56.

[9] Lerner J, Tirole J. Some simple economics of open source. J Ind Econ 2002;50(2):197–234.

[10] Lerner J, Tirole J. The scope of open source licensing. J Law Econ Organ 2005;21(1):20–56.

[11] Stallman R. The GNU operating system and the free software movement. In: DiBona C, Ockman S, Stone M, editors. Open sources: voices from the open source revolution. 1st ed. Sebastopol, US: O'Reilly & Associates, Inc; 1999. p. 53–70.

[12] Free Software Foundation. The GNU general public license. 2013; GNU Project. [Retrieved 18.05.13], from: http://www.gnu.org/licenses/gpl.html.

[13] Bruns B. Open sourcing nanotechnology research and development: issues and opportunities. Nanotechnology 2001;12(3):198–210.

[14] Perens B. The open source definition. In: DiBona C, Ockman S, Stone M, editors. Open sources: voices from the open source revolution. 1st ed. Sebastopol, US: O'Reilly & Associates, Inc; 1999. p. 171–88.

[15] TAPR. The TAPR open hardware license. 2013. http://www.tapr.org/ohl.html.

[16] Chumby HDK, chumby HDK license agreement 2013. Available from: http://www.chumby.com/developers/agreement.

[17] Seidle N. IP obesity. Available from: 2012. http://www.sparkfun.com/news/963.

[18] The Open Source Hardware Association. OSHW community survey. 2012; Available from: http://www.oshwa.org/oshw-community-survey-2012/#section4.

[19] Zelenika I, Pearce JM. The internet and other ICTs as tools and catalysts for sustainable development: innovation for 21st century. Inf Dev, in press. http://dx.doi.org/10.1177/0266666912465742.

[20] Brewer Eric, Demmer Michael, Du Bowei, Ho Melissa, Kam Matthew, Nedevschi Sergiu, et al. The case for technology in developing regions. Computer 2005;38(6):25–38.

[21] James J. Low-cost information technology in developing countries: current opportunities and emerging possibilities. Habitat Int 2002;26(1):21–31.

[22] The open source hardware association. Definition. Available from: http://www.oshwa.org/definition/.

[23] Boldrin M, Levine DK. Against intellectual monopoly. Cambridge: Cambridge University Press; 2008.

[24] Mowery DC, Nelson RR, Sampat BN, Ziedonis AA. The growth of patenting and licensing by U.S. universities: an assessment of the effects of the Bayh–Dole act of 1980. Res Policy 2001;30:99–119.

[25] Garfinkel SL, Stallman RM, Kapor M. Why patents are bad for software. In: Ludlow P, editor. High noon on the electronic front: conceptual issues in cyberspace. Cambridge: MIT; 1999. p. 35–46.

[26] Heller MA, Eisenberg RS. Can patents deter innovation? The anticommons in biomedical research. Science 1998;280:698–701.

[27] Chesbrough H. Open business models: how to thrive in the new innovation landscape. Harvard Business School Press; 2006.

[28] Lemley MA. Patenting nanotechnology. Stanford Law Rev 2005;58:601–30.

[29] Vaidhyanathan S. Nanotechnologies and the law of patents: a collision course. In: Mehta M, Hunt G, editors. Nanotechnology: risk, ethics and law. London: Earthscan; 2006. p. 225–36.

[30] Makker A. The nanotechnology patent thicket and the path to commercialization. South Calif Law Rev 2011;84:1163–403.

[31] Pearce JM. Make nanotechnology research open-source. Nature 2012;491:519–21.

[32] Mushtaq U, Pearce JM. Open source appropriate nanotechnology. In: Maclurcan D, Radywyl N, editors. Nanotechnology and global sustainability. Boca Raton: CRC Press; 2012. p. 191–213.

[33] Stiles AR. Hacking through the thicket: a proposed patent pooling solution to the nanotechnology "Building Block" patent thicket. Drexel Law Rev 2012;4:555–92.

[34] Pearce JM. Open-source nanotechnology: solutions to a modern intellectual property tragedy. Nano Today, in press. http://dx.doi.org/10.1016/j.nantod.2013.04.001.

[35] McGowan D. Legal implications of open-source software. Univ Ill Law Rev 2001;2001(1):241–304.

[36] Weinberg M. It will be awesome if they don't screw it up. Public Knowl 2010; Available from: http://publicknowledge.org/it-will-be-awesome-if-they-dont-screw-it-up.

[37] Bradshaw S, Bowyer A, Haufe P. The intellectual property implications of low-cost 3D printing. ScriptEd 2010;7(1):5–31.

[38] Diamond J. Collapse: how societies choose to fail or succeed. Penguin; 2005, [Revised edition].

[39] Blackburn S. Ethics: a very short introduction. Oxford: Oxford University Press; 2001, p. 101.

CHAPTER 4

Open-Source Microcontrollers for Science: How to Use, Design Automated Equipment With and Troubleshoot

4.1. INTRODUCTION

As we strive to do more and better experiments in less time, much of today's modern scientific equipment demands some level of automation. In the past, automating even the most conceptually simple tasks was a significant undertaking and the time investment was only justified for tasks that needed to be repeated many times (e.g. factory automation of production to make millions of the same product—for example, test tubes). As low-cost basic electronics became more widespread, automation became possible for a widening collection of scientific tasks; however, the skills necessary to create an automated scientific tool were often not worth the investment of a researcher in a much different field. Scientific equipment companies have historically filled the void, which resulted in proprietary automated scientific research tools becoming widely available (e.g. an automated pipetting system). These are the tools that we are all familiar with as most of us have relied on them for our own earlier studies. However, the very fact that these are specialized tools with a relatively small base of customers assures that their costs are very high because of the concomitant overhead associated with traditional manufacturing. These high costs make access to the best-automated scientific tools outside the reach of much of our global collection of scientific labs and at the very least limits even our most well-financed labs to a suboptimal selection of equipment. The open-source hardware movement has provided a solution to these challenges with the rise of a tool that goes by a somewhat peculiar name—the Arduino.

An *Arduino* is an open-source hardware tool – a single-board microcontroller.[1] You can think of it as a light-weight brain behind any scientific

---

[1]The complete plans for the Arduino modules are published under a Creative Commons license (as discussed in detail in Chapter 3). This means you can make your own or if you are an experienced circuit designer, you can develop your own version of the module, extending it and improving it. At the same time if you want to save money from buying a commercially available Arduino, you can make yourself a breadboard version. In the very near future as will be discussed in Chapter 5 on 3-D printing and in Chapter 7, you may be able to print Arduinos in your lab as needed.

59

Open-Source Lab. http://dx.doi.org/10.1016/B978-0-12-410462-4.00004-4

automation-related task. The Arduino was originally developed as an electronics prototyping platform for design students in Italy by Massimo Banzi and David Cuartielles. They had become frustrated by the fact that teaching students the arcane engineering skills necessary to program a conventional microcontroller left little time for their students to focus on design, their real passion. In addition, commercial microcontrollers were obscenely expensive, which greatly limited the number and scope of the projects they could tackle with their students. They fixed both problems. Arduinos cost only US$20-$50 depending on the complexity of your project and you will know how to use them by the time you are finished with this Chapter. At the same time the Arduino team had the foresight to share their innovation using open-source principles that not only allowed all of the rest of us to benefit – but also ensured that they now have a world-wide collaborative team consistently helping make their designs even better (see the Open-source Microcontroller Family below).

The Arduino platform hardware consists of a relatively simple board based on Atmel's ATMEGA8 and ATMEGA168 microcontrollers and on-board input/output support.[2] The software consists of a standard programming language compiler and the boot loader that runs on the board. The software side of the Arduino builds on an earlier open-source software language (wiring) and integrated development environment (IDE) (processing). Thus, the open-source software language (syntax and libraries) is almost identical to C++ with some slight simplifications and modifications, and the processing-based IDE. Overall, the open-source Arduino environment makes it easy to adapt to your particular project as it is both extremely flexible and relatively easy to learn and use for beginners. Without any previous electronics experience, you should be able to get the basics in an hour and get well into a project of your own in an afternoon.

The Arduino board is a lot like your brain—it is extremely powerful and useful—but for it to act on the environment, it needs peripherals (e.g. eyes, legs, hands, etc.). The Arduino can sense the environment by receiving input from a long list of sensors (e.g. chemical, pressure, temperature, light, magnetic, acceleration, ionizing radiation, humidity, electric current and potential, and vibration, etc.) and then based on rules you set affect its surroundings by controlling a similarly long list of outputs (e.g. lights, heaters, motors, and other actuators).

---

[2]The reader should keep in mind that the Arduino is a prototyping platform first and foremost. Thus, it would always be less expensive to design a hardware solution directly around the Amtel chip and, for example, only have the number of inputs and outputs necessary for the project. An example of this is shown in Chapter 5 where we use a Arduino-compatible, simpler and less expensive board to run an open-source 3-D printer. For the most part though when designing scientific equipment under time constraints, the Arduino can act as both a prototype and your working system. The power of the Arduino is in the ease and speed that it enables you to get an interactive tool running at a very low price.

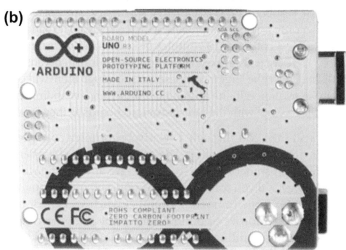

FIGURE 4.1

(a) Front and (b) back of the Arduino Uno board. The Arduino Uno has 14 digital input/output pins (six of which can be used as pulse-width modulation (PWM) outputs, which can be used for controlling power to devices), six analog inputs, a 16 MHz ceramic resonator, a USB connection, a power jack, an In-Circuit Serial Programming (ICSP) header, and a reset button. It contains everything needed to support the microcontroller; simply connect it to a computer with a USB cable or power it with an AC-to-DC adapter or battery to get started.

The Arduino can act as a stand-alone mini computer running experiments or be connected to your laptop via *Universal Serial Bus (*USB) communicating with software running on it acting more like a data logger. As the Arduino is open-source hardware, you can download[3] the complete plans for the board and all the software and make your own. This may be the least expensive route depending on your access to electronics fabrication equipment; however, for most working scientists, it makes more sense to simply purchase the boards preassembled. Enormous libraries of software covering most any kind of sensor or actuator that you will need can be downloaded for free.[4] Although you are strongly encouraged to consider moving your research-related computing to an appropriate open-source operating system[5] (e.g. Debian GNU/Linux[6]) to take advantage of the open-source paradigm discussed in Chapter 2, the Arduino software is truly cross-platform and runs not only on free and open-source GNU/Linux, but also on Windows and the Macintosh OSX operating systems.

As with most open-source hardware projects, the Arduino quickly began a biological-like evolution and now there is an extensive Arduino family that is summarized in the next section along with the other main open-source microcontrollers as of this writing.[7] For some applications, these more powerful tools may be necessary. For example, the Raspberry Pi[8] is essentially a small very low-cost ($35–40) Linux-based computer only missing peripherals, which can greatly expand the ability of a project to handle a collection of complex tasks demanding impressive levels of computation. Here we will primarily focus on the Arduinos because they are the most established, have the largest user base, and although there will certainly continue to be rapid evolution in the Arduino family, by understanding the basics you will also be able to work with the more advanced boards as they are developed.

This chapter will take you through a relatively complex example project for creating an environmental chamber. The environmental chamber is an ambitious first project for someone who does not have any experience with microcontrollers. However, if you follow the steps outlined in Sections 4.3–4.5, you will have a functional environmental chamber when you are done. However, if you want to alter, improve or build on the basic setup described, you will need to have an understanding of C++ and a more in-depth understanding of microcontrollers. At the end of this chapter, several books were recommended

---

[3]The Arduino Project main page: http://arduino.cc/.

[4]Free and open-source Arduino software: http://arduino.cc/en/Main/Software. Free and open-source Arduino libraries: http://arduino.cc/en/Reference/Libraries.

[5]An additional advantage to this approach is that running Linux on an old computer can often "resurrect" it with high-enough performance to be useful for running lab equipment. Most institutions discard computers relatively regularly and you may have free access to all the computing power you need to operate the automated controls for your lab.

[6]Debian: http://www.debian.org/.

[7]Be aware of course that as with all things of open source, the innovation rate is high and thus better designs and new products are constantly appearing.

[8]The Raspberry Pi is a credit card-sized open-source board that plugs into a TV and a keyboard to make a functional Linux-based computer. http://www.raspberrypi.org/.

for providing a firm foundation in Arduino control; however, Adafruit has excellent tutorial library, which should be useful for getting anyone started.[9]

4.2. THE OPEN-SOURCE MICROCONTROLLER FAMILY

There are more than 16 Arduino-based open-source microcontroller boards. The microcontrollers that would be of most use to the science community are summarized in Table 4.1 including their approximate cost, special features, processor, process speed, analog and digital pins, memory, programming language, programmer, and how to expand their capabilities. In the expansion section, the term "shield" refers to interchangeable add-on modules. This is an extremely important feature made possible by the standard way that connectors are exposed on Arduino boards, allowing the CPU board to be connected to a variety of known shields to tackle specific tasks. Some shields communicate with the Arduino board directly over various pins, but many shields are individually addressable via an I²C serial bus, allowing many shields to be stacked and used in parallel. For a list of available Arduino shields, see the Arduino shield list maintained by Jonathan Oxer.[10]

For example, consider the open-source GPS shield developed by Adafruit Industries shown in Figure 4.2. This shield supports any of the four most popular GPS modules and stores data on a standard DOS-formatted SD flash memory card. For scientists doing field work (e.g. environmental scientists tracking water quality in the Great Lakes) after collecting the data, they plug it into their computer and the plain text files are ready for importing into GRASS[11], Google Earth[12], GPSvisualizer[13], or any other form of Geographic Information System (GIS) software or a spreadsheet. The pins indicated in Figure 4.2(b) are required for the SD card. The GPS module also requires pins connected, but

---

[9]Adafurit's Arduino tutorial http://www.ladyada.net/learn/arduino/. Lady Ada's tutorials are very good, but they tend not to cover much in the process control realm, which will be the domain of most researchers interested in pursuing open-source scientific hardware. For more information on that, I recommend a little book by Emily Gertz and Patrick Di Justo called *Environmental Monitoring with Arduino* (2012) published by O'Reilly and Maker Press.

[10]Arduino Sheild List: http://shieldlist.org/.

[11]*GRASS* is the *Geographic Resources Analysis Support System*, which is a free open-source GIS software package used for geospatial data management and analysis, image processing, graphics/maps production, spatial modeling, and visualization. GRASS is currently used in academic and commercial settings around the world, as well as by many governmental agencies and environmental consulting companies. GRASS is an official project of the Open-Source Geospatial Foundation: http://www.osgeo.org/.

[12]*Google Earth* is a virtual globe, map and geographical information program made available for free of charge to download: http://www.google.com/earth/index.html.

[13]GPS visualizer is a free-of-charge utility found on the web that creates customizable maps and profiles from GPS data (tracklogs & waypoints), addresses, or coordinates: http://www.gpsvisualizer.com/.

Table 4.1 Summary of the Open-Source Microcontroller Family

Board:	Arduino Uno	Arduino Leonardo	Arduino Due	MintDuino	Netduino	Netduino Plus	Raspberry Pi	Beagle Bone
Approximate price	$30	$25	$50	$25	$35	$60	$40	$90
Summary	Current "official" Arduino USB board, driverless USB-to-serial, auto power switching	Somewhat experimental Arduino with HID support for mouse or keyboard emulation	Newest Arduino based on a powerful ARM processor. Packs many new features in a Mega sized form factor.	An Arduino-compatible board you build yourself on a breadboard.	Open-Source microcontroller. Programmed using the .NET/C# programming language. Uses an Arduino layout for shield compatibility.	Open-Source microcontroller. Programmed using the .NET/C# programming language. Uses an Arduino layout for shield compatibility.	Single board Linux computer with video processing and GPIO ports	ARM-based hardware hacker focused Linux board.
Special features	Onboard USB controller	HID emulation, USB, SPI on ISP header	Android ADK support, 2 12bit ADC/DAC, USB host, CAN BUS support	DIY Arduino!	Programmed with .NET micro framework.	Programmed with .NET micro framework; onboard ethernet	HD capable video processor, HDMI and composite outputs, onboard ethernet	Onboard USB host and ethernet
Processor	ATmega328	ATmega32u4	32-bit SAM3X8E ARM Cortex-M3	ATmega328	32-bit AT91SAM-7X512-AU	STMicro 32-bit microcontroller	ARM-1176JZF-S	TI AM3358 ARM Cortex-A8
Processor speed	16MHz	16MHz	84MHz	16MHz	48MHz	168Mhz	700MHz	720MHz
Analog pins	6	12	12	6	6	22 (GPIO—digital or analog)	8 (GPIO—digital and analog)	66 (GPIO—digital and analog)
Digital pins	14 (6 PWM)	20 (7 PWM)	54 (12 PWM)	14 (6 PWM)	14	22 (GPIO—digital or analog)	8 (GPIO—digital and analog)	66 (GPIO—digital and analog)

Memory	SRAM 2KB – EEPROM 1KB	SRAM 2.5KB – EEPROM 1KB	SRAM – 96KB	SRAM 2KB – EEPROM 1KB	128KB code storage	384KB code storage	RAM 512MB	RAM 256MB
Programming language	Arduino/C variant	Arduino/C variant	Arduino/C variant	Arduino/C variant	Microsoft .NET environment	Microsoft .NET environment	Any language supported by a compatible Linux distribution (such as Raspbian or Occidentalis)*	Includes Angstrom Linux on SD card. Any language supported by a compatible Linux distribution (such as Ångström or Ubuntu)*
Programmer	USB, ISP	USB, ISP	USB, ISP	Requires programmer like FTDI Friend	USB	USB	You can run any of the Linux-compatible text editors and IDEs right on the Raspberry Pi.	You can run any of the Linux-compatible text editors and IDEs right on the BeagleBone or use the browser-based Cloud9 IDE from another computer.
Expansion	Shield compatible	Shield compatible	Some shields (3V only)	N/A	Some shields	Some shields	Breakout boards such as the pi plate and pi Cobbler.	Capes

*Including Python, Scratch, Perl, Java, JavaScript/Node, C, C++, and Ruby.

(a)

(b)

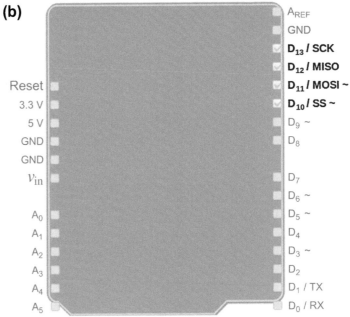

A_{REF}
GND
D_{13} / SCK
D_{12} / MISO
D_{11} / MOSI ~
D_{10} / SS ~
D_9 ~
D_8

Reset
3.3 V
5 V
GND
GND
v_{in}

A_0
A_1
A_2
A_3
A_4
A_5

D_7
D_6 ~
D_5 ~
D_4
D_3 ~
D_2
D_1 / TX
D_0 / RX

FIGURE 4.2

(a) Adafruit Industries open-source GPS shield on an Arduino Uno and its (b) pin diagram.

they can be jumpered to any of the spare Arduino pins. Up to eight other pins may be required depending on whether you implement SD card detection, status indication, and other features.[14]

Next we will walk quickly through the steps to get started with Arduino and then present an example for Arduino-controlled environmental chamber.

4.3.   GETTING STARTED WITH AN ARDUINO MICROCONTROLLER

Very detailed instructions on how to install the Arduino software are provided by the Arduino getting started page for Windows, Mac OSX and many versions of Linux.[15] In keeping with the open-source ethos of this book, we will quickly go through how to install the Arduino software on Linux. First, install Linux on your computer if you are not using it already. The hardest part about doing this is choosing the right flavor of Linux, there are many distributions with various attributes,[16] and I would recommend either going right to Debian[17] or using Linux Mint.[18] You can try Debian by booting a live system from a CD, DVD or USB key without installing any files to your computer. When you are ready, you can run the included installer.[19]

Regardless of what distribution of Linux you chose, you will have access to some form of package maintainer.[20] For example, in Ubuntu Linux, the package manager is called Synamptic Package Manager, which you can get to by search for package manager in the dashboard. Installing the Arduino software is this easy:

1. Open your package manager.
2. Ensure it is updated by clicking "reload".

---

[14]Adafurit's GPS Shield: http://shieldlist.org/adafruit/gps.

[15]Installation instructions: http://arduino.cc/en/Guide/HomePage.

[16]Wikipedia keeps a relatively up-to-date taxonomy: http://en.wikipedia.org/wiki/List_of_Linux_distributions.

[17]Debian is a free operating system (OS) for your computer. An operating system is the set of basic programs and utilities that make your computer run. Debian comes with over 37,500 packages, precompiled software bundled up in a nice format for easy installation on your machine. This makes the installation of for example Arduino absurdly easy. To get Debian, go to http://www.debian.org/distrib/.

[18]Linux Mint is a modern, elegant and comfortable operating system, which is both powerful and easy to use. It is fully free of cost and open source and the fourth most-used home operating system in the world. It is safe and reliable due to a relatively conservative approach to software updates, a unique Update Manager and the robustness of its Linux architecture. Linux Mint requires very little maintenance (no regressions, no antivirus, no anti-spyware... etc.). Linux Mint has better performance than Linux Ubuntu and avoids some of the concerns created by the commercial nature (adware) of Ubuntu's latest releases. To try it out, see http://www.linuxmint.com/download.php.

[19]If you would prefer this "try before you buy" approach, use the link http://www.debian.org/CD/live/ (Note: of course there is no buying, it is all free as in freedom and free as in charge.).

[20]If you are an old-harcore Linux user, you can get it from the terminal "sudo apt-get install arduino".

3. Type in "Arduino".
4. Select the packages you want to install from the list generated (the first two).
5. Click "apply".

The package installer will then (1) install it, (2) put it in your menus and (3) take all dependencies automatically. It just works. Most Windows users will be pleasantly surprised by the ease of the process.

4.4. WORKING WITH THE ARDUINO

Open the Arduino IDE software you installed in the last section. It should look like Figure 4.3.

FIGURE 4.3

Screenshot of Arduino software interface.

The Arduino IDE contains

- a text editor for writing code (including features for cutting/pasting and for searching/replacing text);
- a message area (that gives feedback while saving and exporting and also displays errors);
- a text console, which displays text output by the Arduino environment including complete error messages and other information (also the bottom right-hand corner of the window displays the current board and serial port);
- a toolbar with buttons for common functions;
- a series of menus.

The IDE connects to the Arduino hardware (the board) to upload programs and communicate with them.

When you write software programs in the text editor to control your Arduino microcontroller, they are called *sketches* and saved with the file extension .ino.[21] The toolbar buttons allow you to do the following in order: verify programs (check for errors), upload (compile and upload to your microcontroller) programs, create a new sketch, open of the existing sketches in your sketch-book,[22] and save sketches, and open the serial monitor, which displays serial data being sent from the Arduino board (USB or serial board). The icons are fairly self-explanatory, but are also defined if you hover the mouse above them. Additional commands are found within the five menus[23]: File, Edit, Sketch, Tools, and Help. Sketch also enables you to import libraries, which inserts one or more #include statements at the top of the sketch and compiles the library with your sketch. This is handy but can cause problems if you begin to push the limits of your Arduino's capabilities. Because libraries are uploaded to the board with your sketch, they increase the amount of storage space that is used. If this becomes issues for you and a sketch no longer needs a library, simply delete its #include statements from the top of your code. *Tools* also enable you to select the Arduino board you are using.

The Arduino IDE uses the concept of a *sketchbook*, which is a standard place to store your sketches. Open your sketches in your sketchbook from the File>Sketchbook menu or from the Open button on the toolbar. The first time

---

[21]Early versions of the IDE prior to 1.0 saved sketches with the extension .pde, which you may run across on the Internet—particularly for the most basic sketches. It is possible to open these files with version 1.0 or higher, but you will be prompted to save the sketch with the .ino extension.

[22]The latest version of this menu does not scroll. This can be cumbersome if you have a long list of sketches. So if you need to open a sketch far down on your list, use the File | Sketchbook menu instead.

[23]*Please note that the menus are context sensitive*—so you will only be able to use the items relevant to the work currently being carried. For detailed explanations of each menu item, see http://arduino.cc/en/Guide/Environment.

you run the Arduino software, it will automatically create a directory for your sketchbook. You can view or change the location of the sketchbook location with the Preferences dialog. The software allows you to manage sketches with more than one file with each file showing up handily in its own tab. These files can be of several types including your standard Arduino code files (no extension), C files (.c extension), C++ files (.cpp), or header files (.h).

After you have written your sketch from scratch (or downloaded/altered/customized it) and you want to test it in the real world, you need to upload it to the Arduino board. To do this, you need to ensure that you are using the correct board and serial port. You can check this by going to the Tools > Board and Tools > Serial Port menus, respectively. The correct board is the one you are using (e.g. an Arduino Uno). The board selection has two effects: it sets the parameters (e.g. CPU speed and baud rate) used when compiling and uploading sketches; and sets and the file and fuse settings used by the burn bootloader command. For serial ports, it will depend on your operating system:

- On *Linux*, it should be /dev/ttyUSB0, /dev/ttyUSB1 or similar (e.g. Or /dev/ttyACM* for non-FTDI-USB-to-serial converters).
- On *Windows*, it is probably COM1 or COM2 (for a serial board) or COM4, COM5, COM7, or higher COM10,20 (for a USB board)—to find out, you look for USB serial device in the ports section of the Windows Device Manager.
- On the *Mac*, the serial port is probably something like /dev/tty.usbmodem241 (for an Uno or Mega2560 or Leonardo) or /dev/tty.usbserial-1B1 (for a Duemilanove or earlier USB board), or /dev/tty.USA19QW1b1P1.1 (for a serial board connected with a Keyspan USB-to-Serial adapter).

After selecting the correct serial port[24] and board, press the upload button in the toolbar or you can select *File* > Upload in the menu. Current Arduino boards will reset automatically and begin the upload. With older boards (pre-Diecimila) that lack autoreset, you need to press the reset button on the board just before starting the upload. On most boards, the RX and TX LEDs blink (as seen around the middle of the board to the left of the Arduino logo in Figure 4.1(a)) as the sketch is uploaded. The Arduino environment will display a message when the upload is complete.

When you upload a sketch, you are using the Arduino bootloader, a small program that has been loaded on to the microcontroller on your board. It allows

---

[24]If you are having trouble figuring out the correct port, try unplugging, check the list, plug it back in, and then select the port that was not there when it was unplugged.

you to upload code without using any additional hardware. The bootloader is active for a few seconds when the board resets; then it starts whichever sketch was most recently uploaded to the microcontroller. The bootloader will blink the on-board (pin 13) LED (above the RX and TX LEDs in Figure 4.1(a)) when it starts (i.e. when the board resets). You can also use the IDE to put a boot-loader on a raw microcontroller.

To make the work of programming sketches really easy, I strongly recommend utilizing libraries whenever possible. Libraries provide extra functionality for use in sketches. For example, they are used when working with hardware or manipulating data. The open-source Arduino programming community has probably already-written libraries for most of the individual mini-programs you need to write to get your own piece of open-source hardware working. To use a library in a sketch, select it from the Sketch > Import Library menu. This will insert one or more #include statements at the top of the sketch and compile the library with your sketch. For a list of libraries, see http://arduino.cc/en/Reference/Libraries. Some of the most commonly used libraries are included with the Arduino software, but the global collection of open-source libraries is continually expanding and also being developed for other Arduino-compatible boards (e.g. Sanguino, Lira, TinyDuino, etc.). If your research demands installing non-Arduino libraries, see this very straightforward guide.[25] As you research group becomes more adept at using Arduinos, you may want to develop your own libraries and as discussed in Chapter 2, share them to benefit from the global open-source support. For step-by-step instructions, see the Arduino hacking tutorial.[26]

4.5. EXAMPLE: THE "POLAR BEAR" OPEN-SOURCE ENVIRONMENTAL CHAMBER

Arduino microcontrollers can be used with a combination of relatively inexpensive supplements available in any hardware store or Wal-Mart to replace very sophisticated and expensive research tools. Consider the research-grade environmental chamber. An environmental chamber is simply an enclosed volume, in which the temperature and humidity are controlled. They are required for many experiments where it is important to control the environment in disciplines as far removed as biology to civil engineering to materials science. Our group was interested in the effects of temperature and humidity on the solidification and mechanical properties (e.g. interlayer adhesion)

---

[25]Library Installation Guide: http://arduino.cc/en/Guide/Libraries.
[26]Arduino hacking tutorial: http://arduino.cc/en/Hacking/LibraryTutorial.

of various polymers undergoing an extrusion process.[27] We looked online for commercial environmental chamber systems that would be appropriate (both humidity and temperature range and the necessary volume to fit a RepRap 3-D printer comfortably). Our requirements were modest and flexible, but the lowest cost commercial solution was still over US$6500. This seemed unnecessary considering the functionality of the device so we decided to make our own. The design that follows highlights some of the functionality of the Arduino including the ability to both measure and produce an output based on that measurement.

In addition, the open-source environmental chamber showcases the Arduino's ability to solve research-related instrumentation challenges for a small fraction of the cost of commercialized solutions. It is unlikely that you will want to use our environmental chamber design for the same purpose as our lab, but if you need a similar environmental chamber—perhaps a larger one with a greater temperature range—the underlying code and design we have developed is easily adapted. This again is the power of open source, you or anyone else will have a much easier time designing and building an environmental chamber after finishing reading this section and having access to our code. As many people begin to use this design, improve it, and feed it back into the community; the next time my lab needs an environmental chamber, I am quite confident we will have access to a superior designs (e.g. higher performance for less cost).

This system was designed and built in less than a semester by Rodrigo Faria, an undergraduate student researcher in my group with assistance with other members of the group and the online open-source hardware community. The "Polar Bear" Open-source Environmental Chamber obtained its name from the distinctive polar bear ultrasonic humidifier,[28] which is used to increase humidity when necessary within the chamber as seen in Figure 4.4(a). The complete wiring diagram is shown in Figure 4.4(b) along with the pin map of the Arduino Mega 2560 in Figure 4.4(c), which is the brain of the Polar Bear Chamber [1]. If you are inexperienced in electronics, do not be intimidated by this diagram, pin map or the spaghetti-bowl of wires in the picture. The system may appear extremely complex, and it is, but in the Wiring Diagrams subsection, we will walk through how to wire each individual component separately along with their function. In this way, you will gain a complete understanding of the system, while at the same time cutting

---

[27]This interest stemmed from the fact we observed radical fluctuations in print quality using the open-source 3-D printers discussed in Chapter 5 depending on the time of year and the equally radical (and sometimes uncomfortable) environmental conditions in our laboratory.

[28]There is nothing in particular special about the polar bear other than it is cute and was the least expensive viable humidifier solution we found during the time of the design. If it really is important to you and you do not like the polar bear, you can make your humidifier look like anything else you can CAD using 3-D printing described in Chapter 5.

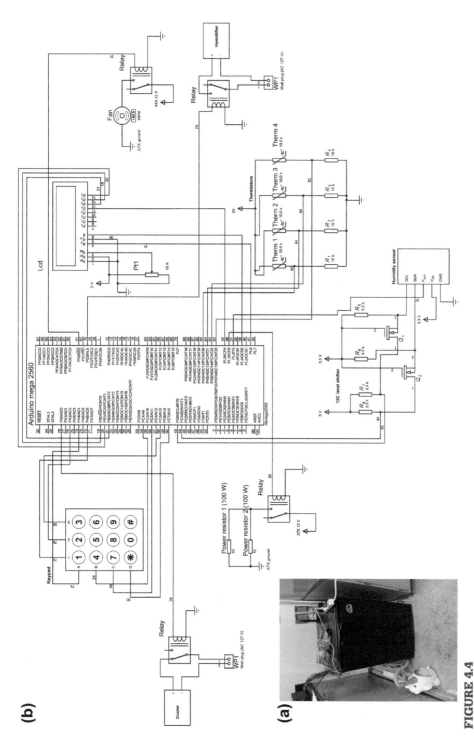

FIGURE 4.4

(a) The physical and (b) the complete wiring diagram of the "Polar Bear" Open-source Environmental Chamber and (c) the pin map of the Arduino Mega 2560, which is the brain of the Polar Bear Chamber.

(c)

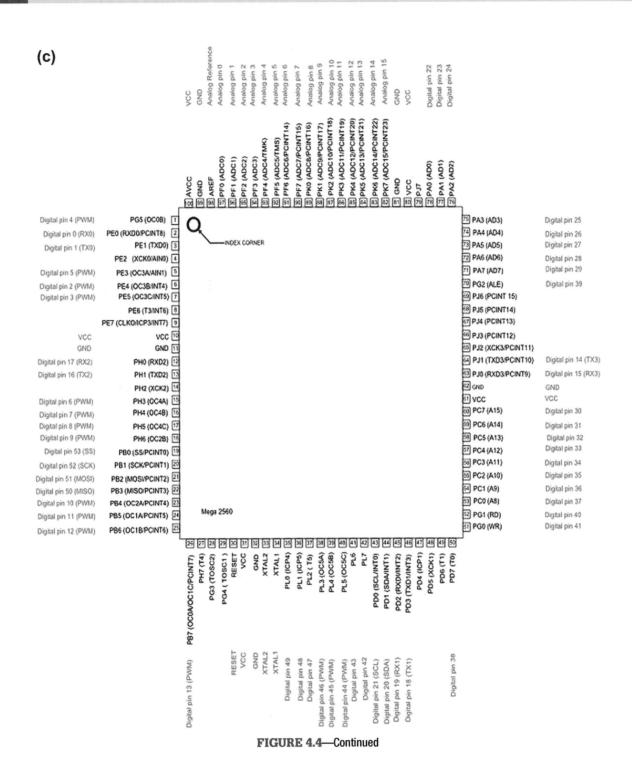

FIGURE 4.4—Continued

through the complexity. Figure 4.4(b) and (c) is meant primarily as a reference as you build to double check that you have attached all the components correctly.

4.5.1. Specifications

The "Polar Bear" Open-source Environmental Chamber (Bear Chamber) was designed to provide the necessary environmental control at the lowest possible cost in the shortest amount of time. The chamber is designed around the shell of a dorm refrigerator and thus has a work area volume of over 100 l (3.65 cubic feet). It can control the chamber relative humidity between 10% and 90% and the temperature between the range 25 °C below and 25 °C above the ambient temperature. The power draw on the Bear Chamber was measured with a Watt meter during operation and is summarized in Table 4.2.

Table 4.2 Measured Power Consumption of the Polar Bear Open-Source Environmental Chamber

Setting	Power Consumption [W]
Idle	12.5
Heating	48.6
Cooling	115
Dehumidifying	54
Humidifying	67.5
Estimated overall use with chamber closed	57

4.5.2. Bill of materials polar bear open-source environmental chamber

The list of components, source, number and approximate cost is shown in Table 4.3. A few points should be made about the prices listed in Table 4.3. As with most laboratory equipment builds, a certain quantity of salvaged components was utilized. For example, in our case, we had available metal sheets, wires, power supplies and computer fans from our in-lab "store" of recycled components. Most organizations are constantly discarding old computers. Even very old computers can gain a new life using a fast and light-weight Linux operating system; however, even those that are too old to be used as equipment interfaces can be harvested for their power supplies, heat sinks, wires, fans and case materials. For this project, for example, salvaged computer components saved about $50 on our total cost, excluding shipping costs. The prices for purchased components are correct at the time of design for the vendor listed. When possible, several alternative vendors are also shown. Table 4.3 is also

Table 4.3 Bill of Materials for the Polar Bear Open-source Environmental Chamber

Component	Source(s)	Number	Approximate Cost (US$)
Arduino mega 2560	Arduino, adafruit, Sparkfun, etc.	1	$42.00
9 V power adapter for Arduino (1000 mA)	Adafruit http://adafruit.com/products/63	1	$6.95
5 V relays (JZC-11F-05VDC – 1Z)	Sparkfun https://www.sparkfun.com/products/100	4	$7.80
Humidity sensor (HH10D)	Sparkfun https://www.sparkfun.com/products/10239	1	$9.95
10 kΩ thermistors	Sparkfun https://www.sparkfun.com/products/250	4	$7.80
16 × 2 character LCD 5 V	Sparkfun https://www.sparkfun.com/products/255	1	$13.95
12-button keypad	Sparkfun https://www.sparkfun.com/products/8653	1	$3.95
Connection wires	Electronics shop, web	Several feet	$2.50
100 W power resistors (resistance value: 10 Ω, power rating: 100 W)	Electronics shop, web	2	$12.58
3.3 kΩ resistors	Electronics shop, web	4	$0.25
10 kΩ resistors	Electronics shop, web	4	$0.25
10 kΩ potentiometer or trimpot	Electronics shop, web	1	$0.95
ZVN4424A MOSFETs (N-channel enhancement mode vertical DMOS FET)	Electronics shop, web	2	$2.10
Ultrasonic humidifier (Sunpentown SU-2031 polar bear ultrasonic humidifier)	Amazon, web	1	$32.00
Renewable wireless mini dehumidifiers (Eva-dry E–500)	Amazon, web	1-3	$49.89
12 V fan	Or from a old used computer		$5.00
12 V DC power supply (24 A current drain capable)	Salvaged computer	1	$30
Refrigerator/cooler	Wal-Mart, web	1	$117.00
Hose	Hardware store, web	1	$5.00
Waterproof silicone adhesive	Hardware store, web	1	$5.00
Metal sheets (any metal with good heat transfer)	Hardware store, web	2	$10
		Total	~$365

grouped by vendor in order to try to constrain shipping costs. For those with mature labs, already many of the components like wiring and simple circuit elements will already be available for trivial or no costs. As you can see from Table 4.3, the bottom line is the Polar Bear Chamber which can be built for less than $400 in materials.

The Polar Bear Chamber is made up of eight subsystems, all controlled by the Arduino, which will be individually detailed in wiring diagrams below:

- The LCD interface, which enables feedback to the system operator on the state of the system.
- The keyboard interface, which enables the system operator to input desired environmental conditions and run the Arduino sketch.
- The thermistors that measure the chamber temperature.
- The humidity sensor that measures the relative humidity within the chamber.
- The power resistors that are used to heat the chamber when the chamber temperature is below the temperature set point.
- The Polar Bear humidifier, which is used to increase the humidity when it is below the humidity set point.
- The refrigerator, which is used to cool the chamber when it is above the temperature set point.
- The fan, which is used to blow humid air over the passive dehumidifier when the chamber humidity is above the humidity set point.

4.5.3. Wiring diagrams

For experienced electrical engineers, the wiring and circuits in the Polar Bear Open-source Environmental Chamber are pretty simple; however, to most researchers, the inside of the finished product is going to look like a bowl of spaghetti. To cut through the complexity and make it easy for everyone, the wiring diagram is described below and shown separately for each of the eight subsystems that make up the Polar Bear. Wiring each subsystem is fairly straightforward. As you get near the end, simply ignore what is going on in your bowl of spaghetti wires and just make sure the circuit you are constructing matches the wiring diagram for the subsystem.

The wiring diagram for the LCD display to the Arduino is shown in Figure 4.5 and needs the Arduino, potentiometer, and the LCD display along with connection wires.

The following nine connections should be made as seen schematically in Figure 4.5:

1. Gnd→Potentiometer—Arduino Ground
2. V_{cc}→Potentiometer—Arduino 5 V
3. V_o→Potentiometer sliding contact
4. R_s→Digital Pin 43
5. E→Digital Pin 45

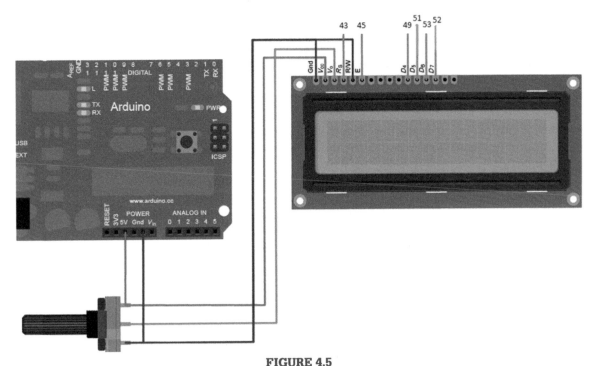

FIGURE 4.5
Wiring diagram for the LCD display to the Arduino.

6. $D_4 \rightarrow$ Digital Pin 49
7. $D_5 \rightarrow$ Digital Pin 51
8. $D_6 \rightarrow$ Digital Pin 53
9. $D_7 \rightarrow$ Digital Pin 52.

It should be noted that Figure 4.5 shows the schematic for an Arduino Uno, and the code we used to interface with the LCD was originally written for it. This again shows the ease of porting and adapting previous open-source hardware investment and code to your own projects.

The wiring diagram for the keypad is shown in Figure 4.6. For this subassembly, only the Arduino and keypad are needed. As can be seen in Figure 4.6, the pins to be connected are Arduino Digital Pins 24, 26, 28, 30, 32, 34 and 36.

In order to obtain a good global temperature estimate inside the chamber, four thermistors are used in the Polar Bear. The circuit diagram for the thermistors is shown in Figure 4.7. The circuit shown in Figure 4.7 must be made for each thermistor. Connect the thermistor to the Arduino 5 V output and a 10 kΩ as a voltage divider and then connect the other resistor's side to the ground pin. A wire must be connected between the 10 kΩ resistor and

FIGURE 4.6

Wiring diagram for the keypad.

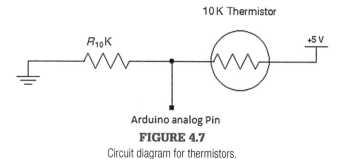

FIGURE 4.7

Circuit diagram for thermistors.

the thermistor; this wire will be connected to the Arduino Analog Pins. Each thermistor must be connected to a single-analog pin; the pins are A_{12}–A_{15}.

The humidity sensor (HH10D) needs to be wired as shown in Figure 4.8. It is connected to the Aurdino and the MOSFET (ZVN4424A).

However, as both the HH10D [2] and Arduino operate at different voltage values, the humidity sensor must be wired through a Logic Level Shifter circuit as shown in Figure 4.9 [3].

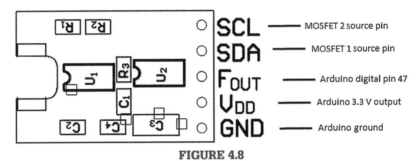

FIGURE 4.8
Wiring diagram for the humidity sensor.

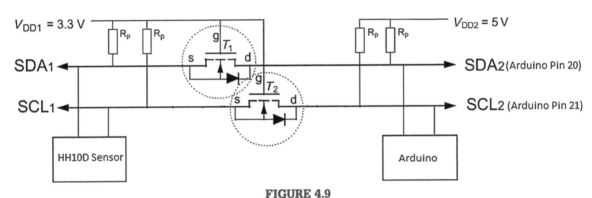

FIGURE 4.9
Wiring diagram for level shift circuit.

The resistors "R_p" in Figure 4.9 have values of 3.3 kΩ. V_{DD1} and V_{DD2} values can be found in the Arduino board. It should be noted that any N-Channel Enhancement Mode Vertical DMOS FET can be used if it operates on the following ranges:

- Gate threshold voltage min 0.1 V to max 2 V
- On resistance max 100 Ω at I_D of 3 mA and V_{GS} of 2.5 V
- Input capacitance max of 100 pF at V_{DS} of 1 V and V_{GS} of 0 V
- Switching times of max 50 ns
- Allowed drain current of 10 mA or higher.

The two power resistors are turned on and off by the Arduino that controls a relay to the 24 A power supply. Connect the power resistors as shown in Figure 4.10. Be sure to connect the relay to *Arduino Digital Pin 48* and the power resistors to the *Normally Opened pin* on the relay.

Similar to the power resistors, the humidifier is connected to the Arduino through a relay as shown in Figure 4.11. The humidifier AC plug cable has two wires, and one must be split. Connect one end of the split to the *Normally Opened pin* on the relay and the other to the common pin. Set the humidifier knob to medium and connect to a wall plug. Connect the relay to *Arduino Digital Pin 39*.

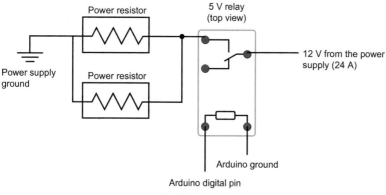

FIGURE 4.10

Wiring diagram for the power resistors.

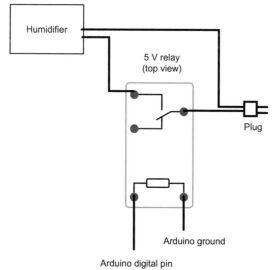

FIGURE 4.11

Wiring diagram for the humidifier.

Next, connect the fan wires properly respecting the polarity as shown in Figure 4.12. Connect the relay to *Arduino Digital Pin 41*.

For cooling in the prototype Bear Chamber we used a dorm refrigerator, but depending on your application you may want a smaller or larger chamber. The chamber will be made up by the size of your refrigerator or cooler. Refrigerator/coolers usually have a thermostat that is connected to their main protection relay. To find it, you will need the cooler's wiring diagram or manual. If you

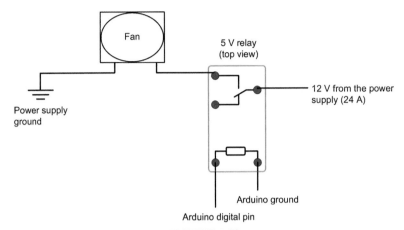

FIGURE 4.12

Wiring diagram for the fan.

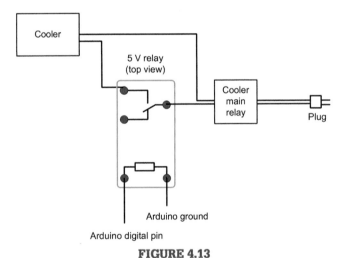

FIGURE 4.13

Wiring diagram for the refrigerator/cooler.

buy it new, obtaining the wiring diagram is not a problem. Similarly, you can normally find manuals for relatively recent appliances on line. If it is an older salvaged refrigerator, you may need to contact the manufacturer to get it. Once you have their diagram, detach their thermostat and connect the main protection wires to the *Normally Opened Pin*[29] on the relay as shown in Figure 4.13.

---

[29]All those normally opened pins are on a different relay each.

Connect the Relay to *Arduino Digital Pin 22*. In this way you are essentially controlling a relay with the Arduino that controls the Coolers main relay. Then plug the refrigerator or cooler as you would normally into an AC wall outlet when you are ready to proceed.

It is always useful to have a wiring organization. To do so, as you build the system, think carefully about soldering boards, the use of zip ties, and apply your creativity. Sometimes it is also helpful—particularly for wire runs—to dry run them first before making final length and soldering decisions.

4.5.4. Construction of the Bear Chamber

The construction of the Bear Chamber is relatively simple even if the untidy state of a nonenclosed system is a mess of wires in their "naked" state as seen in Figure 4.14.[30] The key step is to find the correct wires for the cooler protection relay attached to the thermostat, remove the thermostat and connect the wires to the Arduino relay as instructed above.

The construction of the Bear Chamber begins with preparing the cooling chamber. In this example case, we gutted the inside of the dorm refrigerator to make space for a Mendel Prusa style RepRap like the 3-D printer described in the next chapter. Next, drill a hole from outside to the inside of your chamber to run the following wires: power resistors, fan, and thermistors. This "data and power" hole should be near the top of the chamber. Next, use a sturdy sheet (we used plywood, but this will depend on your application), cut to size to split the cooler in two sections—the upper section contains most of the parts and also is the work area; the bottom section will be used to place the dehumidifier(s). The base sheet must be thick enough to mechanically support whatever you want to put in your environmental chamber. It must have a square hole cut in it for the fan and some small holes for air circulation. This can be done most easily with an electrical drill and a jigsaw—but can be accomplished by hand tools or whatever you have available (e.g. laser cutter, water jet cutter, etc.). Next, fix the metal sheets on the cooler sides together with the power resistors as seen in Figures 4.15 and 4.16.

Then run the wires for the thermistors into the chamber and place the thermistors in different, but relevant areas such as near (but not on) a power resistor, top back, top front and bottom again roughly following Figure 4.15. These can be held in place with duct tape (as shown, with the advantage of offering quick alterations) or can be bolted, fastened, or glued in place. Next, place

---

[30]If aesthetics are important to your lab, you can house the control board and wiring in a simple box that can sit on top of your chamber. Similarly if you do not like the look of the bear, you can choose a different humidifier or box that as well.

FIGURE 4.14
Polar Bear open-source Environmental Chamber with "naked" wiring.

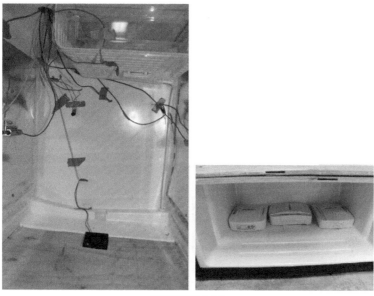

FIGURE 4.15

Inside of the Bear Chamber showing placement of components.

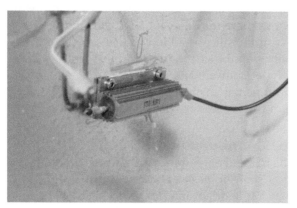

FIGURE 4.16

Detail of mounting of the power resistor on the metal plate.

the humidity sensor somewhere in the back of the chamber far from power resistors. Finally, place the dehumidifiers underneath the base sheet in their mini-chamber as seen in Figure 4.15. The chamber is now ready to control the temperature and extract humidity.

FIGURE 4.17
Hose connections between the chamber and the polar bear dehumidifier.

The last step involves finishing the ability to add humidity. Drill a second hole near the bottom of the chamber in your "work zone" for the humidifier entry port. The hole must have the diameter large enough to fit the hose. Insert the hose and seal with the silicone adhesive; also seal the other end of the hose on the humidifier with the silicone as shown in Figure 4.17. Your Bear Chamber is now ready for software and operation. Congratulations you just made a >$6000 environmental chamber for less than $400! Make sure everything is plugged in and you are ready to go.

4.5.5. Bear Chamber software and controls

To download the free and open-source software needed to run your Bear Chamber, go to http://sourceforge.net/projects/bearproject/.[31] Download the file Enviro_Chamber_V1_RodrigoFaria.zip by clicking on the green download button. This software written by Rodrigo Faria should work out of the box. It will not involve any changes or coding understanding if you followed the wiring instructions above. However, improvements, improved functionality, or alternative implementations can be done (and are encouraged) by assigning different pins to the components and by editing the programming source code.[32]

---

[31]SourceFourge.net is a free and open-source web-based source code repository. It provides fast downloads from the largest open-source applications and software directory on the Internet.
[32]To do this, some understanding of C programming and the Arduino board are required.

Inside the zip file are two folders: (1) Enviro_Chamber_V1_RodrigoFaria and (2) libraries along with a readme.txt file. The first folder contains the Arduino .ino files specific for this project and the second, the generalizable libraries that make coding with an Arduino so fast and easy. Follow the instructions in the readme.txt, which are the following:

1. Copy the folders inside libraries to Arduino's libraries folder on your computer.
2. Open Enviro_Chmaber_V1_RodrigoFaria and upload to Arduino MEGA 2560.[33] To do this, follow the details in the "Working with Arduino" section above. Be sure to select the correct serial port and board, press the upload button in the toolbar or select *File > Upload* in the menu.

4.5.6. Controller concepts

With the sketches uploaded to the Arduino, the Bear Chamber is now ready for operation and is controlled with the keypad and display as shown in Figure 4.18. There are two basic control concepts that need to be understood to operate the Bear Chamber: temperature and humidity.

After you set the desired temperature (T in °C) and the controller to ON, the controller will acquire the actual temperature from the sensors and compare with the desired temperature. Depending on the result, the system will enter a heating or cooling stage, where the power resistors or the cooler will be activated. This is shown schematically in Figure 4.19.

When the temperature reaches a satisfactory range, it will be maintained by activating the heaters or the cooler only when necessary. This is just good energy management, but is critical depending on your application (e.g. if you are using the Bear Chamber in the field powered with a solar photovoltaic system). A satisfactory temperature range for our application is from 0.5 °C below the desired temperature to the set point (desired value); in another words, if the desired temperature is T, the range is $[T-0.5\,°C;\,T]$.

The humidity controller is similar to the temperature controller. After you set the desired relative humidity (H in %) and set the controller to ON, the controller will acquire the actual relative humidity inside the chamber from the sensor and compare with the relative humidity wanted. Again, depending on the result, the system will enter a humidifying or drying stage, where the humidifier or the fan will be activated. This is shown schematically in Figure 4.20.

---

[33]In addition, I would suggest that you copy the other part into the sketchbook, which makes it easier to open the file, and keeps your hard disk organized.

If dehumidification is necessary, the fan will be activated. Once the fan is activated, the air will be forced to circulate to the bottom of the chamber where the dehumidifiers will dry the air. The dehumidification process is the slowest of the control processes. This must be factored into experiments, especially when lower values are set. When the relative humidity

FIGURE 4.18

Details of "naked" Bear Chamber control board and wiring.

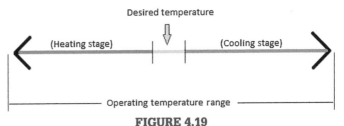

FIGURE 4.19

Operating temperature range of Bear Chamber.

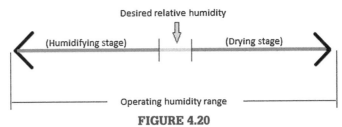

FIGURE 4.20

Operating relative humidity range of Bear Chamber.

reaches a satisfactory range, it will be kept by activating the humidifier or the fan just when necessary, again contributing to good energy management. A satisfactory relative range is from 3% below to 3% above the desired relative humidity (set point); in another words, if the desired relative humidity is H, the range is [H − 3%; H + 3%].

4.5.7. Operation of the Bear Chamber

The operation of the Bear Chamber is easy and straightforward. When the system is turned on, the main screen will appear as shown in Figure 4.21.

The main screen will show the temperature in degree Celsius, relative humidity in percentage and controller status. The possible commands are further information by pressing "*" on the keypad or controller menu by pressing "#" on the keypad.

The further information menu "*" will show the temperature and relative humidity set points (desired values) defined by the user (by default, temperature is set to 25 °C and relative humidity to 50%) as seen in Figure 4.22.

The available command on this screen is "*", which will step you to the next information screen and show you the relative humidity set point. Finally, in the relative humidity set point screen, press "*" to go back to the main screen (Figure 4.21).

Once you are back in the main screen (Figure 4.21), if # is pressed, the controller menu will be shown. The available commands are "1"—Set Controller Status; "2"—Set Temperature; "3"—Set Relative Humidity; or "*" to go back to main screen. In *Controller Status* screen, press "1" to activate controller and "2" to deactivate controller or "*" to go back to the main screen. On the temperature screen,

FIGURE 4.21

Main screen of Polar Bear open-source environmental chamber.

FIGURE 4.22

Temperature set point screen under further information.

FIGURE 4.23

Control screen to input the temperature set point in degrees celsius into the Bear Chamber.

FIGURE 4.24

Control screen showing target temperature confirmation.

press "1" to set temperature or "*" to exit to the main screen. When "1" is pressed, the user will be prompted to enter a desired temperature as shown in Figure 4.23.

After you have entered your desired temperature, a check screen will be shown (Figure 4.24).

If the temperature entered is correct, press "#" to accept or "*" to edit. Once the temperature is accepted, the system will show the main screen (Figure 4.21). If the controller is already ON, the system will automatically start; otherwise, set the controller to ON in order to have the system operating. The menu controls for the relative humidity information, input, and confirmation screens and set point process are identical to the temperature. After setting up the temperature once, this should be intuitive.

4.5.8. Troubleshooting

Similar to experiments, rarely does scientific hardware builds work perfectly the very first time. Anticipate that your lab will need to spend some time on troubleshooting—particularly if you are trying new tools or microcontrollers, your group has not built up a specialty in yet. Hopefully, however, the troubleshooting will be fast and easy. If your project is having general problems with the Arduino (e.g. you get errors when launching the Arduino software, you cannot upload a sketch, or you are having problems getting your sketch to run), see the Arduino Trouble Shooting Guide.[34] It is an excellent step-by-step guide to solving all the most common types of problems with the microcontroller itself.

---

[34]Arduino Trouble Shooting Guide: http://arduino.cc/en/Guide/troubleshooting.

For the Bear Chamber specifically, there are a few common problems that are easily fixed:

- *The screen is too dark or you cannot see anything on the screen:* Make sure the Arduino is connected to the power adapter and rotate the potentiometer near the LCD screen to adjust the brightness.
- *After increasing the relative humidity and trying to reduce the humidity afterwards, the relative humidity takes a prohibitively long time to change:* As discussed above, it is always going to take longer to dehumidify than humidify in this chamber. Experiments in general should thus be scheduled appropriately (e.g. if you are doing tests as a function of humidity start with the least humid test first). If that is not possible, the best option is to open the door to restart at your room humidity. So if the system achieved a relative humidity above 40% and the user wants something lower than 20%, open the system's door and wait until the excess of moisture is released and then close. Do not open the chamber frequently if low-humidity conditions are demanded.
- *The screen is showing weird letters:* Restart the system; to avoid this error, good isolation and proper wire connection are required. We have found that there can be interference from power supplies, so it is recommended that you locate the power supply away from the controls circuitry.
- *The display shows incorrect or strange relative humidity readings (e.g. negative values):* Check for equipment capable of emitting radiofrequencies near the system and remove them if possible. The power supply can cause readings errors; so keep it as far away from the controls as practical. If the problem persists, implement a band pass filter for the relative humidity sensor's frequency output (i.e. frequency band: 5–10 KHz).
- *The keypad does not respond or is working improperly:* Some keypads have different wiring setups, so it is possible that your wiring may be slightly different than the examples shown in this book. You need to find the correct pins to be assigned. Follow this Arduino tutorial,[35] and then connect the pins following Figure 4.25.
- *The relative humidity does not increase when programmed to do so:* Verify that the humidifier water tank has water in it.
- *The relative humidity does not decrease:* Verify that the dehumidifiers are still able to absorb moisture by looking at the front of your Eva-Dry as shown in Figure 4.26. If the color inside the rectangle is near orange, it needs to be recharged; to do so, remove from the Bear Chamber and plug into the wall until it turns blue again. Do not leave it plugged in for more than 18 h as the manufacturer warns this can cause a malfunction.

---

[35]Arduino keypad tutorial: http://playground.arduino.cc//Main/KeypadTutorial.

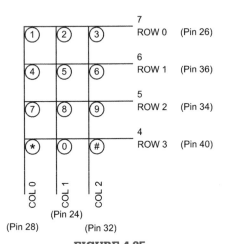

FIGURE 4.25
Troubleshooting pin connection for keypad.

FIGURE 4.26
Color indicator on dehumidifier.

4.5.9.   Open-source technological evolution for the Bear Chamber

If you build the Polar Bear Open-source Environmental Chamber following the instructions above, it will work and perform to the specifications. This, however, may not be adequate for your environmental chamber needs. Perhaps you need a larger chamber or one that can cool to $-10\,°C$. Similarly, depending upon your technical expertise, you may have many ideas for ways to improve the design, function, code, etc. of the Bear Chamber. Excellent! Please do it. This piece of scientific hardware is open source [4], which means that you can use all the designs for free to make your laboratory operate better for less money, but we expect in return that you feed your improvements back into the open-source ecosystem.

If you make improvements, share your source files and instructions with everyone else. Depending on your improvement, it may be appropriate to share it with:

- The Arduino.cc community.
- The code at sourceforge.net/projects/bearproject/.
- The Appropedia project page: http://www.appropedia.org/Open-source_environmental_chamber.

4.6.   CONCLUDING THOUGHTS AND ADDITIONAL READING

In this chapter, we have covered the basics of open-source microcontroller use for scientific equipment and detailed an example where the open-source paradigm creates a research-grade tool for less than 1/16th of the cost of commercial equipment. As we will see in Chapters 5 and 6, this is not an anomaly. Open-source hardware, in general, costs much less than commercial hardware and is customized to your labs use. Even if your research never is in the need of an environmental chamber, walking through the exercise above should have prepared you to use Arduinos or similar open-source microcontrollers to design, use and troubleshoot automated scientific tools of your own if you know C++. If you would like to learn more about Arduino microcontrollers and really pick up the basics from the first step, in addition to the project website, some of these texts may be useful to you:

- Banzi, M. (2011). *Getting Started with Arduino. 2nd Edition.* Make. (This is an excellent guide to provide to graduate students to take them from no experience to being useful for working with Arduinos to make scientific equipment).
- Margolis, M. (2011). *Arduino Cookbook.* O'Reilly Media, Incorporated. (This provides many example projects, written more for the Maker community that may be of use on some scientific projects).
- Monk, S. (2012). *Programming Arduino: Getting Started with Sketches* McGraw-Hill. (This is a slightly more advanced text on programming for more complicated projects.)

REFERENCES

[1] Arduino. Pin map of the Arduino Mega 2560. Available from: http://arduino.cc/en/uploads/Hacking/PinMap2560.png.

[2] HH10D Datasheet; 2013. http://www.sparkfun.com/datasheets/Sensors/Temperature/HH10D.pdf.

[3] Philips Semiconductors. Bi-directional level shifter for I²C-bus and other systems. August, 2007. http://ics.nxp.com/support/documents/interface/pdf/an97055.pdf.

[4] Pearce JM. Building research equipment with free, open-source hardware. Science 2012; 337(6100):1303–4.

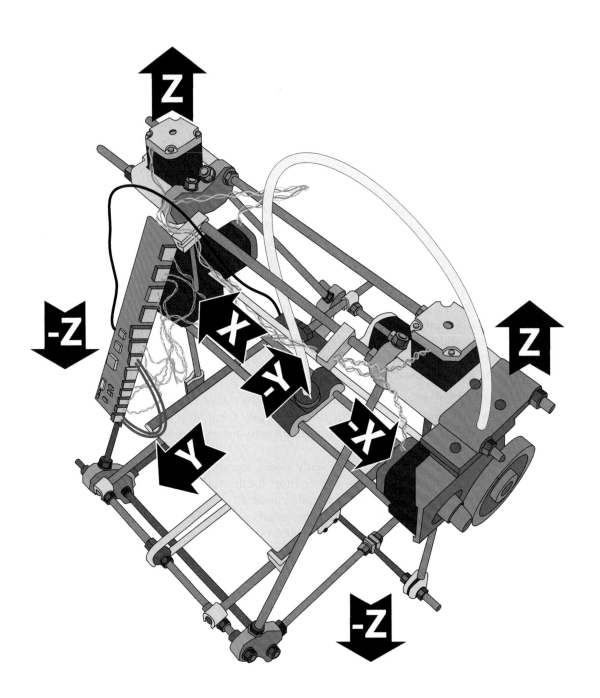

RepRap for Science—How to Use, Design, and Troubleshoot the Self-Replicating 3-D Printer

5.1. INTRODUCTION TO REPRAPS

Although the term 3-D printing only goes back to 1995, the concept of 3-D printing or, as it is known in the technological literature, "additive manufacturing" of AM has been around for a while used for rapid prototyping since the late 1980s. The more recent technical explosion in innovation for 3-D printing has been substantial, fueling rapid growth in commercial rapid prototyping for a huge swath of industries [1–5]. There has been a considerable amount of hype built up around 3-D printing at this point and even speculation by the *Economist* that these technical advances could result in a "third industrial revolution" governed by mass-customization and digital manufacturing following traditional business paradigms [6]. In the *Economist* special issue, however, lost among all the fascinating stories of new companies and breathless explanations of the technologies is the fact that *open-source 3-D printing* is really what has cut the cost of rapid prototyping so low that it is now accessible to basically everyone. The recent development of open-source 3-D printers makes the scaling of mass-distributed additive manufacturing of high-value objects technically feasible for everyone [7–15]. Yet perhaps the first group to benefit the most financially is practicing experimental scientists [16]. These self-*rep*licating *rap*id prototypers, or RepRaps were developed by Dr Adrian Bowyer, a professor in mechanical engineering at the University of Bath in 2005. Bowyer has a particularly clear view[1] of what having a general-purpose self-replicating manufacturing machine means for humanity and how the project was able to literally evolve under the open-source paradigm [17,18]. This is why the major variants of the RepRap are all named after biologists Darwin, Mendel, Huxley, and Wallace. The RepRap evolutionary family tree [19] is shown in Figure 5.1.

---

[1]In Wealth Without Money, Bowyer writes: "The self-copying rapid-prototyping machine will allow people to manufacture for themselves many of the things they want, including the machine that does the manufacturing. It is the first technology that we can have that will simultaneously make people more wealthy, while reducing the need for industrial production…" [18].

Open-Source Lab. http://dx.doi.org/10.1016/B978-0-12-410462-4.00005-6

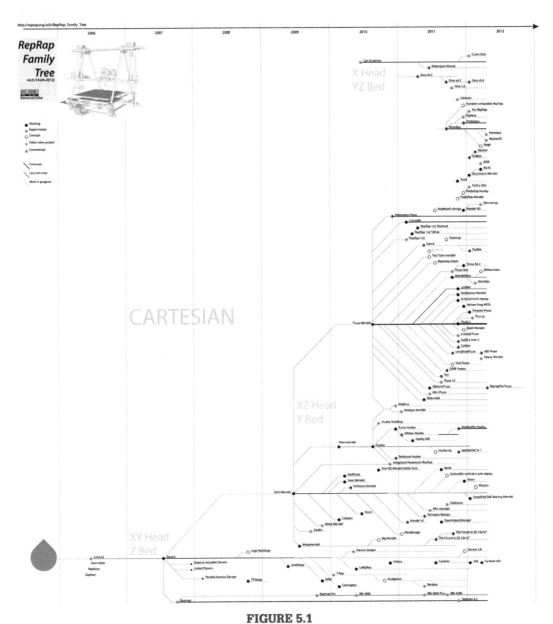

FIGURE 5.1

The RepRap Family Tree—Timeline 2006-2012 v4.0 14-04-2012.

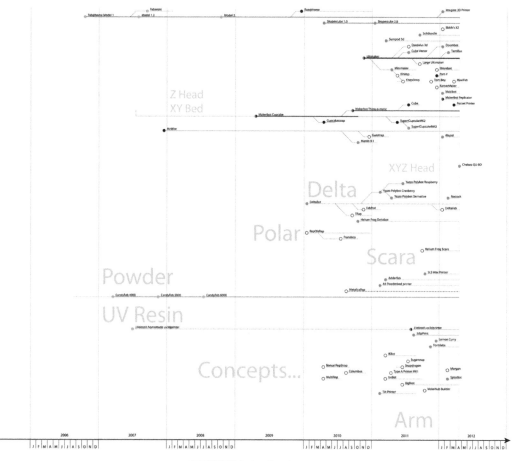

FIGURE 5.1—Continued

The MOST RepRap this chapter discusses in detail is a Mendel Prusa variant as shown in Figure 5.2.

The RepRap project has created 3-D printing machines that can manufacture approximately half of their own parts (57% self-replicating potential excluding fasteners, bolts and nuts) from sequential fused deposition of a range of polymers and common hardware [7,20,21]. The RepRap is a mechatronic device made up of a combination of printed mechanical components, stepper motors for 3-D motion and extrusion, and a hot-end for melting and depositing sequential layers of polymers, all of which is controlled by an open-source microcontroller such as the Arduino discussed in the last chapter [7,22,23]. The extruder intakes a filament of the working material (polylactic acid (PLA),

FIGURE 5.2

A new variant of the Prusa Mendel RepRap and open-source 3-D printer capable of fabricating about half of its own parts. In the picture, all the translucent blue parts were printed on an identical mechatronic machine. The variants were engineered by G. Anzalone and this particular printer was built by J. Irwin. At this point, the printer shown is a great-great-great-great-great granddaughter of one of Bowyer's original RepRaps.

acrylonitrile butadiene styrene (ABS), and high-density polyethylene among materials [24]), melts it using resistive heating, and extrudes it through a nozzle. The printer uses a three-coordinate system, where each axis involves a stepper motor that makes the axis move until reaching a limit switch for homing. The printing process is a sequential layer deposition where the extruder nozzle deposits a 2-D layer of the working material, then the z (vertical)-axis will raise, and the extruder will deposit another layer on top of the first building a 3-D object in 2-D layers. RepRaps have been proposed and demonstrated to be useful for developing engineering prototypes [20], education [25], creating electronic sensors [26,27], cocreative product realization [28], personal manufacturing [29], modular robotics [30], wire embedding [31], tissue engineering [32] and appropriate technology manufacturing for sustainable development [11]. Here, we are, of course, most interested in the RepRap as an extremely powerful and low-cost tool for helping to fabricate customized scientific equipment [16,26]. In the sections that follow, we will discuss how to build a RepRap, and use it to make open-source scientific hardware.

5.2. BUILDING A REPRAP

The version of the RepRap detailed in this chapter is a variant of the Prusa Mendel reengineered primarily by Jerry Anzalone at Michigan Tech. We have built several of them at Tech, including several pilot workshops where dozens of teachers (college and high school) were trained in how to build and debug them. The goal of this particular version is a low-cost RepRap with high print speeds, good quality and that it is extremely easy to build and debug. For this purpose, the MOST RepRap has a new *y*-axis stage that gives you both a larger effective print envelope and the ability to increase print speed. If you follow the instructions below, your RepRap will be ready in <24 h of work (two people can build it comfortably in a working 8 h day, particularly if they are experienced). That said, just like all RepRaps, these designs are constantly evolving and the most up-to-date build instructions can be found in the Appropedia wiki.[2] My research group maintains a small community centered on an Appropedia page to assist in troubleshooting and share user experiences.[3] In addition, there is a much larger RepRap Forum[4] and community at the main RepRap wiki.[5]

Before assembling, first ensure you have all the parts necessary in the bill of materials (BOM) as shown in Table 5.1, the printed parts in Table 5.2, and the necessary tools are listed below. First download the zip file of the MOST STereoLithography (STLs).[6] Convince someone you know[7] with a 3-D printer to print all of the STLs shown in Table 5.2 with PLA unless otherwise indicated by ABS following the file name (you can print them all in ABS, but it is unnecessary and in general, PLA is better to work with because of substantially less fumes). Printing all the parts on a standard RepRap printer with normal settings should take <24 h of print time. The printed parts for the MOST RepRap in the link provided are plated and color coded for easy identification and assembly. In addition, if you simply print all the plates you can be sure your part count is right in place of Table 5.2.

The following tools will make the building of your RepRap fast and easy although they are not all absolutely necessary. For example, you do not absolutely need work gloves, but then you need to be more careful handling the edges of sharp

---

[2]MOST RepRap Build http://www.appropedia.org/MOST_RepRap_build For the two person build use: http://www.appropedia.org/MOST_RepRap_parallel_build_overvie.
[3]MOST RepRap User Experience http://www.appropedia.org/MOST_HS_RepRap_user_experience.
[4]RepRap Forums http://forums.reprap.org/.
[5]RepRap wiki http://www.reprap.org.
[6]MOST RepRap Printable component STL files—http://www.thingiverse.com/thing:92321.
[7]If you do not know anyone with a 3-D printer and do not have any other form of rapid-prototyping or 3-D printing services at your institution, there are numerous commercial 3-D printing services available along with individuals in the RepRap forums that will print the components out for you for a nominal fee. For example, see http://forums.reprap.org/list.php?175.

Table 5.1 3-D Printer Bill of Materials (BOM)

		Part Number/ SKU	Source	Units	Unit Cost	Total
1	Controller	1284P	http://matterfy.com/products/melzi-ardentissimo-1284p	1	$120.00	$120.00
2	6T5 timing belt	B6T5-MPS	http://shop.polybelt.com/6T5-Open-End-Belt-Roll-Polyurethane-with-Steel-Cords-B6T5-MPS.htm	7	$1.09	$7.63
3	608 bearings	608zz	http://www.amazon.com/VXB-Skateboard-Bearings-Double-Shielded/dp/B002BBGTK6/ref=sr_1_3?ie=UTF8&qid=1351727094&sr=8-3&keywords=608zz	2	9.47	$18.94
4	Limit switch	SW767-ND	http://www.digikey.com/product-detail/en/SS-3GLPT/SW767-ND/664728	3	$1.11	$3.33
5	Thermistor for heated bed	495-2157-ND	http://www.digikey.com/product-detail/en/B57891M0103J000/495-2157-ND/739907	1	$0.83	$0.83
6	Heated build platform	817752010412	http://www.lulzbot.com/en/14-heated-print-bed.html	1	$33.00	$33.00
7	Hot end	817752010047	http://www.lulzbot.com/en/166-budaschnozzle-11.html	1	$95.00	$95.00
8	Hobbed bolt	817752012416	http://www.lulzbot.com/en/7-hobbed-bolts.html	1	$7.00	$7.00
9	M8 metric drill rod	88625K67	http://www.mcmaster.com/#drill-rods/=jyro7o	3	$5.59	$16.77
10	M8 metric DIN 125 18-8 SS flat washer	93475A270	http://www.mcmaster.com/#metric-flat-washers/=jyovoj	1	$7.90	$7.90
11	M6 metric 18-8 stainless steel hex nut	91828A251	http://www.mcmaster.com/#metric-hex-nuts/=krvyhs	1	$8.73	$8.73
12	M3 metric DIN 125 18-8 SS flat washer	93475A210	http://www.mcmaster.com/#metric-flat-washers/=jyp0sb	1	$1.62	$1.62
13	M4 washers	93475A230	http://www.mcmaster.com/#metric-flat-washers/=jys2cq	0.08	1.86	$0.15
14	M8 metric 18-8 stainless steel hex nut	91828A410	http://www.mcmaster.com/#metric-hex-nuts/=jyov6p	2	$9.98	$19.96
15	M3 metric DIN 125 18-8 SS nut	91828A211	http://www.mcmaster.com/#metric-hex-nuts/=jyp186	1	$5.55	$5.55

Table 5.1 3-D Printer Bill of Materials (BOM) *Continued*

		Part Number/ SKU	Source	Units	Unit Cost	Total
16	M4 metric DIN 934 18-8 SS nut	91828A231	http://www.mcmaster.com/#metric-hex-nuts/=jys1i0	0.04	6.45	$0.26
17	M3 × 8 metric 18-8 SS cup point set screw	92015A105	http://www.mcmaster.com/#metric-set-screws/=jypdj2	0.04	$6.56	$0.26
18	M3 × 20 metric 18-8 SS socket head cap screw	91292A123	http://www.mcmaster.com/#metric-socket-head-cap-screws/=jyoylf	1	$5.74	$5.74
19	M3 × 10 metric 18-8 SS Socket head cap screw	91292A113	http://www.mcmaster.com/#metric-socket-head-cap-screws/=jyoylf	0.06	$5.85	$0.35
20	Metric 18-8 SS nylon-insert hex locknut	93625A300	http://www.mcmaster.com/#nylock-nuts/=jyrz5m	0.02	9.98	$0.20
21	Hex cap M4 × 50	91287A057	http://www.mcmaster.com/#standard-cap-screws/=jys0xf	0.16	5.78	$0.92
22	M8 metric 18-8 stainless steel threaded rod	90024A080	http://www.mcmaster.com/#standard-threaded-rods/=jyoueu	6	$8.31	$49.86
23	Springs	9657K272	http://www.mcmaster.com/#compression-springs/=krvzrx	1	$9.58	$9.58
24	1/8 × 1/4 Polytetrafluoroethylene (PTFE) tubing	5033K31	http://www.mcmaster.com/#standard-ptfe-tubing/=krvxad	2	$3.73	$7.46
25	LM8UU linear bearings	14003979	http://www.suntekstore.com/goods-14003979-6pcs_lm8uu_8mm_linear_ball_bearing_bush_bushing.html	2	$7.13	$14.26
26	Stepper motor	UMN17MTR	http://ultimachine.com/content/kysan-1124090-nema-17-stepper-motor	5	$16.50	$82.50
27	Carbon fiber kite rod	20961	http://www.goodwinds.com/merch/list.shtml?cat=carbon.pultrudedcarbon	1	7.79	$7.79
28	Glass		Local		0.99	$0.00
29	Power supply, 12V/20A	9SIA0U008P5040	http://www.newegg.com/Product/Product.aspx?Item=9SIA0U008P5040&nm_mc=KNC-GoogleMKP&cm_mmc=KNC-GoogleMKP-_-pla-_-NA-_-NA	1	27.99	$27.99

Continued

Table 5.1 3-D Printer Bill of Materials (BOM) *Continued*

	Part Number/ SKU	Source	Units	Unit Cost	Total
Total					$553.58
Single Purchase					
30 Heat shrink tubing	344	https://www.adafruit.com/products/344		$4.95	
31 Solder	734	https://www.adafruit.com/products/734		$24.95	
32 Flux	779008835661	http://www.amazon.com/dp/B0080X79HG/ ref=biss_dp_t_asn		$8.95	
33 Wire ties	GB 50098	http://www.amazon.com/GB-50098-Electrical-Assorted-500-Pack/dp/ B00004WLJ9/ref=sr_1_5?ie=UTF8&qid=-1351724946&sr=8-5&keywords=wire+ties		$9.55	
34 Lubricants	Local				
35 Filament	Various	http://ultimachine.com		Varies	

components. Similarly, there are other alternative tools you could use to clean out holes of printed components, but the sizes lists will make it a faster task.

List of recommended tools:

1. Wrenches: (2) 13 mm, 7 mm, and 5.5 mm.
2. Hex key (Allen key) set: 5, 4, 3, 2.5, 2, and 1.5 mm.
3. Measuring tape.
4. Needle-nose pliers.
5. Utility knife blades—used as scrapers to clean the sharp edges of printed parts.
6. Drill bit size 5/16″ and #9 (0.196″) (to clean out holes; ideally 8, 5 and 3 mm).
7. Wood jigs (290 and 234 mm) shown in Figure 5.3.
8. Gloves—to prevent cutting fingers on threaded rod and sharp edges of printed parts.
9. White lithium grease (to lubricate bearings).
10. Torpedo or bubble level.
11. File or thread-chaser to clean up cut threaded rods.
12. Zip ties for wire management.
13. Small vice grips to hold drill bits for reaming (or you can use 3-D printable handles)[8]
14. Pencil sharpener—if you do not have one, you can use a knife or utility knife blades.
15. Silver epoxy.
16. Wire strippers.

[8]Drill bit handles http://www.thingiverse.com/thing:91035.

Table 5.2 RepRap Printed Parts

Part Name	Part Images
Frame vertex foot1 4×	
Frame vertex 1 2×	
Hearingbone gears	

Continued

Table 5.2 RepRap Printed Parts *Continued*

Part Name	Part Images
Twelve tooth T5 gear1 2×	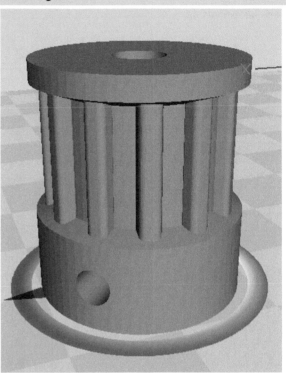
Buda mount	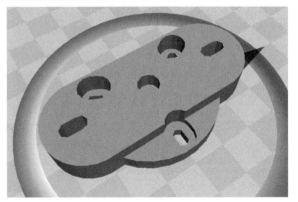

Table 5.2 RepRap Printed Parts *Continued*

Part Name	Part Images
Prusa extruder mount	
z-motor mount 2x	
Bar clamp 8×	

Continued

Table 5.2 RepRap Printed Parts *Continued*

Part Name	Part Images
2× melzie mount	
Parametric coupling 2x	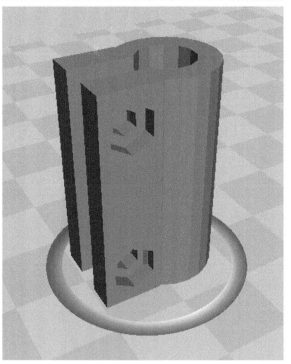

Table 5.2 RepRap Printed Parts *Continued*

Part Name	Part Images
C_rod_Y_axis	
Wades plate	
X_end_plate	

Continued

Table 5.2 RepRap Printed Parts *Continued*

Part Name	Part Images
y-motor bracket	
Belt clamp 2x	
Belt terminator 4×	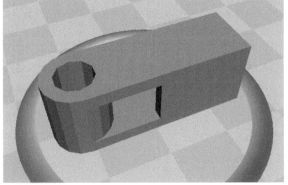

Table 5.2 RepRap Printed Parts *Continued*

Part Name	Part Images
Prusa extruder mount	
z-mounter mount2	
Wireholder 2x	

Continued

Table 5.2 RepRap Printed Parts *Continued*	
Part Name	**Part Images**
3× end stop holder	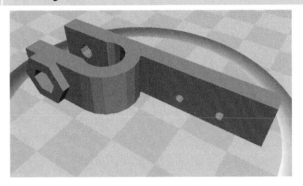

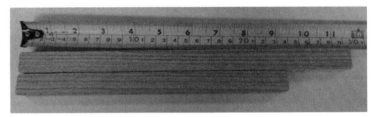

FIGURE 5.3

Wood jigs for measuring RepRap frame.

5.2.1. Bill of materials by component

In order to assist in fast assembly, the BOM from Table 5.1 is broken down here into main components. The complete BOM is shown in Figure 5.4. I recommend purchasing the entire BOM and then breaking them up and ordering in the follow subcategories, which we will then step through building.

5.2.1.1. Frame

Printed parts

Four frame-vertex-foot.stl

Two frame-vertex.stl

Two bar-clamp.stl

Steel

Six 370 mm M8 threaded rod

Four 294 mm M8 threaded rod

Three 440 mm M8 threaded rod

Sixty-six M8 nut

Seventy M8 washer

FIGURE 5.4
The complete BOM for the MOST RepRap.

5.2.1.2.   z-axis

Printed parts
    Two z-motor-mount.stl
    Two parametric-coupling.stl
    Two bar-clamp.stl
    One end-stop-holder.stl
    One rod-clamp.stl
Steel
    Four LM8UU bearing
    Two 350 mm 8 mm smooth rod
    Two 210 mm M8 threaded rod
    Ten M8 nut
    Sixteen M8 washer
    Seven M3 × 20mm screw
    Eight M3 × 10mm screw
    Fifteen M3 washers
    Seven M3 nut
    Two M2 × 14mm screw
    Two M2 washer
Electromechanical
    Two NEMA17 1.8° stepper motor
    One Limit switch

5.2.1.3.   x-axis

Printed parts
    1×_end_plate.stl
    1× carriage.stl (ABS)

One belt-clamp.stl
Two belt_terminator.stl
One end-stop-holder.stl
One 12t_T5_pulley.stl

Steel

Two 608ZZ bearing
Two 385 × 8mm smooth rod
One 50 × 8mm threaded rod
Two M8 nut
Two M8 washer
Two M8 fender washer
Three M3 × 25mm screw
Three M3 × 20mm screw
Five M3 × 10mm screw
Nine M3 nut
Nine M3 washer
Two M2 × 14mm screw
Two M2 washer

Electromechanical

One NEMA17 1.8° stepper motor
One Limit switch

5.2.1.4.   y-axis

Printed parts

One Y-Motor_bracket.stl
Two belt_terminator.stl
One end-stop-holder.stl
One 12t_T5_pulley.stl
One y_carriage.stl (ABS)

Steel

Four 608ZZ bearing
Four LM8UU bearing
Four 420 × 8mm smooth rod
Twelve M8 nut
Fourteen M8 washer
Thirteen M3 × 10mm screw
Five M3 × 20mm screw
Fifteen M3 nut
Thirty-two M3 washer
Two M2 × 14mm screw
Two M2 washer

Electromechanical

One NEMA17 1.8° stepper motor
One Limit switch

5.2.1.5. Extruder

Printed parts

    One wades_plate.stl

    One hearingbone_gears.stl

    One exdrive_spacer_direct.stl (ABS)

    One spacer_w_insert.stl (ABS)

    One prusa_extruder_mount.stl

    One buda_mount.stl (ABS)

Steel

    One M8 or 8 mm smooth rod, 25 mm long for idler

    Four M4 × 50 hex cap

    Ten M4 washers

    One hobbed bolt

    Four springs

    Six M4 nuts

    Two M4 × 25 mm screws

    One 8 mm nut and washer

    Three 608 bearings

    Four M3 × 10 mm screws

    Four M3 washers

    One M3 × 10mm grub screw

    One M3 nut

    Two M6 nut

Electromechanical

    One NEMA17 1.8° stepper motor

5.2.2. Assembly instructions

There are some exceptionally detailed and well-written descriptions about how to build a standard Prusa RepRap. For the simplified version of the RepRap presented here, rather than tell you exactly where to place each nut and bolt, figures are provided that should make it apparent from the list of parts for that particular component. We have tried these build instructions out with several faculty and student builds, which were successful with only minor supervision from experienced RepRap builders. You should be able to build the RepRap on your own, but it is more easily accomplished with a team of two. Ideally, you would have access to an experienced RepRap user to help you through any unclear parts. Even if there is no one in your company or institution available, the RepRap Internet Relay Chat (IRC) channel is always available to let the open-source community answer quick questions.[9]

---

[9]http://reprap.org/wiki/IRC. It should be noted that this is an anonymous Internet forum and just like any other anonymous forum, you should not take offense or "feed the trolls" that sometimes interject less than useful comments. The majority of RepRap IRC users are experienced and helpful.

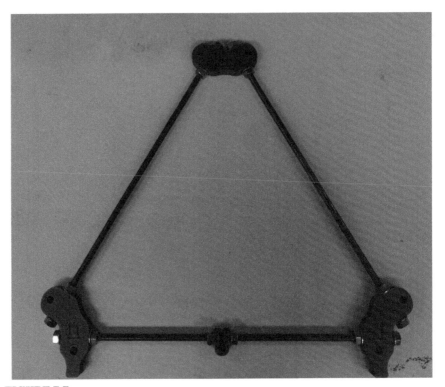

FIGURE 5.5
The equilateral triangle on each side of the frame of the MOST RepRap.

5.2.3. The frame

There is an equilateral triangle on each side of the frame of the MOST RepRap as seen in Figure 5.5. Initial assembly will be very loose. Do not tighten the nuts or be overly concerned about their placement during first assembly steps; everything will be tightened up and properly placed in future steps.

Follow the image in Figure 5.5 and build two triangles each with two footed frame vertices, one unfooted frame vertex, one rod clamp on the threaded rod connecting the two footed vertices, 14 M8 nuts, 14 M8 washers and three 370 mm M8 threaded rods.[10] Using the 290-mm-long jig, space each vertex from its neighbors and tighten the nuts so that a rigid triangle is formed. The jig should be placed such that it is against the printed vertices near the washers, but not on the washers. To make sure the holes line up, you can lay the triangles on top of each other and insert six M8 bolts through each hole in the

---

[10]Make sure you have the feet facing the correct direction at the bottom of the triangle.

FIGURE 5.6
The assembled MOST RepRap frame.

vertices. The bolts should go through holes in both vertices easily, adjust the frame a little if necessary.

5.2.3.1. Building the bulk of the frame

When you build this, make sure you do not tighten down any of the bolts—it should be super loose to start. The additional parts needed to fabricate the rest of the frame with your two triangles are three 440 mm M8 threaded rods (the long ones), 16 M8 washers and 16 M8 nuts.

Put the frame together with the long rods as shown in Figure 5.6. You are already done with the frame! Nice work.

5.2.4. The y-axis (back and forth)

The y-axis consists of two sides each consisting of a pair of threaded rods with only slight differences in what is placed on these rods. The assembly of the top rods is identical, whereas the lower rods are assembled slightly differently. Note the distinction between motor end and idler end—the motor end has additional nuts and washers on the lower rod.

FIGURE 5.7
The proper order of assembly for *y*-axis upper rods.

5.2.4.1.   Building the y-axis upper rods

Sometimes the rods are hard to thread. If you are having difficulty thread-ing, chase the threads with an M8×1.25 die or jam a pair of nuts on one end of the threaded rod and use wrenches to increase the amount of torque applied.

You will need two sets of the following parts needed for each side:

Ten M8 washers
Two 608 bearings
Eight M8 nuts
M8 threaded rod of 294 mm
Two rod clamps

Starting from the left order of assembly (assembly is symmetric about the 608 bearings): washer, nut, nut, washer, rod clamp, washer, nut, nut, washer, washer, 608 bearing, 608 bearing, washer, washer, nut, nut, washer, rod clamp, washer, nut, nut, and washer. The proper order is shown in Figure 5.7.

Build two identical copies of Figure 5.7. Mount them through the upper holes of the bottom, footed vertices. Add a washer and nut to each end, but continue to leave the assembly loose to accommodate placement of parts during later steps.

5.2.4.2.   Building the bottom rods

The parts you need for motor end are the following:

Two 294 mm M8 threaded rod.
Six M8 washers.
Six M8 nuts.

Parts needed for idler end are the following:

Two 294 mm M8 threaded rod.
Four M8 washers.
Four M8 nuts.

FIGURE 5.8

Assembly of the motor end, bottom rod (motor side) for the y-axis.

FIGURE 5.9

Assembly of the idler end bottom rod (idler side) for the y-axis.

To assemble on the motor end, bottom rod (motor side) follows Figure 5.8: washer, nut, nut, washer, washer, nut, nut, and washer.

Then follow Figure 5.9 for the idler end bottom rod (idler side): washer, nut, nut, and washer. Mount rods to the frame with a washer and a nut on each side are shown in Figure 5.10. It does not matter which side at this point of the build. Again, remember to leave the assembly loose at this point—tightening will take place later.

5.2.5. The z-axis (up and down)

First, obtain all the z-axis components shown in Figure 5.11.

To assemble the z-motor mounts, face the y motor mount toward yourself and then on the right put on z-motor mount with a washer, washer, and nut. Use the smaller jig to get the distance between the two triangles more or less right. Then put on the second z-motor mount on the left with washer, washer, and nut. Mount the motors to the motor mount with M3 screws as shown in Figure 5.12. Next, put a nut, washer, rod clamp, washer, and nut on each side of the bottom threaded rod. Then put the vertical smooth rod through the bottom-rod clamp and use a level to adjust the bottom-rod clamps to the correct position. Lubricate the LM8UU bearings with white lithium grease. Attach the smooth rod from bottom to top on both sides with two bearings on each smooth rod as seen in Figure 5.12.

5.2.6. The x-axis (left–right)

First, obtain all the x-axis components shown in Figure 5.13. Then, assuming that the printed parts you are working with are not perfect, clean up motor end

FIGURE 5.10
Mounted *y*-axis rods to the frame.

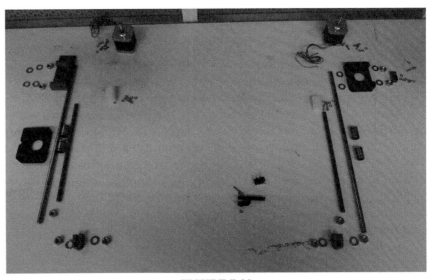

FIGURE 5.11
Components of *z*-axis.

FIGURE 5.12

Motors installed on the motor mounts and smooth rods installed for the z-axis.

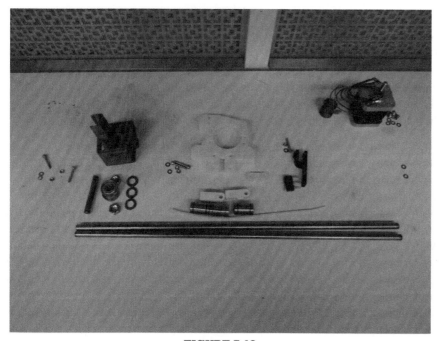

FIGURE 5.13

Components of x-axis.

FIGURE 5.14

Motor end of the *x*-axis before (right) and after (left) cleaning with a knife or needle-nose pliers.

of the *x*-axis with a knife or needle-nose pliers as seen in the yellow parts of Figure 5.14, which show before (right) and after cleaning (left). You also need to use a knife to dig out a hole in the big nut trap, but using a drill bit is usually easier. Load M3 nuts into the *x*-axis and use a threaded rod to push through to the bottom and then thread the screw for both of the deep nut traps as shown in Figure 5.15.

Next construct the idler as shown in Figure 5.16 in the following order: nut, washer, plastic, washer 2 bears, washer, and nut. After the idler is complete, load bearings on the motor and idler sides (two on each of the *z*-axis) and mount on vertical smooth rods as shown in Figure 5.17. Slide the two smooth rods through both of the plastic *x*-axis sides. Mount the *x*-stage with three bearings on the two smooth *x*-rods (two bearings on the motor side and one on the other). Then put M3 × 20s all the way through the idler side with washers and bolts. For details, see Figure 5.17. Your *x*-axis should be complete so now make sure it is level. You can also use a plumb bob to ensure that the vertical smooth rods are directly vertical as shown in Figure 5.18. Repeat this step on the other side. The stage should slide up and down effortlessly, when it does tighten it all down. You should now have a completed frame with *x*, *y* and *z*, all set up as shown in Figure 5.19.

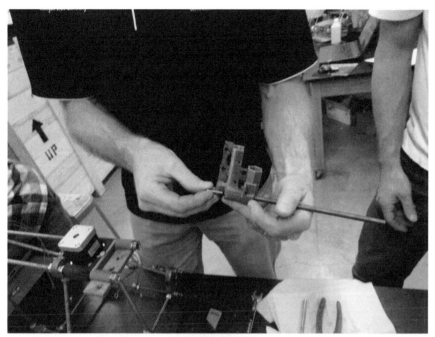

FIGURE 5.15
Loading the deep nut traps.

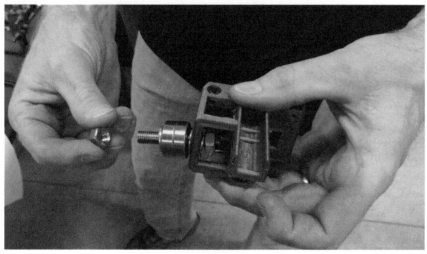

FIGURE 5.16
The idler of the x-axis.

FIGURE 5.17
Loaded bearings and mounted on vertical smooth rods of the *x*-axis.

The last step to complete the main portion of the mechanical assembly is to attach the *x*-stage to *z*-stage. Take the couplings that couple the shaft to the motor and clean them out. These couplings take a pair of M3 × 20 screws and M3 nut and M3 washers and they fit into the nut traps on the coupler. Mount the coupler and the *z*-axis threaded rod (210 mm M8 threaded rod) as seen in Figure 5.20. Please note that the couplers are printed in ABS for flexibility. The coupling ends may need to be reamed out with a drill bit, however, ensure that there is a snug fit after reaming and prior to tightening the screws.[11] Pull the

---

[11]Make sure you have added the nuts and springs to the threaded rod before putting on the *x*-axis.

FIGURE 5.18

Using a plumb bob to ensure that the vertical smooth rods are directly vertical.

x-axis up—use a wire to temporarily tie it to the top frame and put two M8 nuts on the *z*-axis threaded rod on both sides.

Release the *x*-axis back onto the nuts as shown in Figure 5.21. Use three small zip ties to tie the *x*-axis stage to the three bearings. Next, mount the pulley to the *x*-axis stepper motor with M3 trap nuts and M3 × 10 grub screws and mount the motor with three M3 × 10 screws as shown in Figure 5.22. Cut 875 mm of belt and attach the belt terminator as seen in Figure 5.23. Pull on the belt

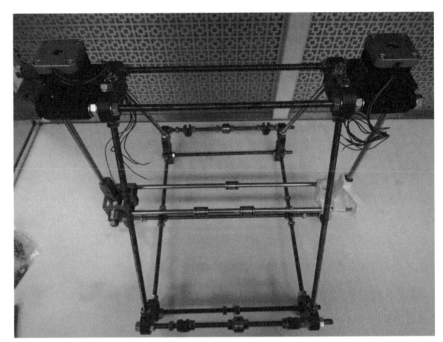

FIGURE 5.19

A completed frame with *x*, *y* and *z*.

carefully and the teeth will mate. This may take a couple tries until you get it to lock in tightly. Run the belt through the pulley on the motor side and then the idler and put on the second belt terminator using the same method shown in Figure 5.23. Then use a zip tie to tie both terminators together.

The terminators should be on the top as shown in Figure 5.24. Use a belt clamp and two M3 × 20 nuts, washers and bolts to attach the belt to the *x*-axis stage. While attaching, make sure that the belt terminators and the stage are on opposite sides of the printer (as shown in Figure 5.24).

5.2.7. Attaching the y-stage

Slide out the two smooth rods for the bottom *y*-stage and put on two LMU bearings on each side. Make sure the smooth rods are on top. Snap on the *y*-motor mount in between the double sets of washers on the base (Figure 5.25). Mount the pulley with two M3 nuts and M3 grub screws to the last stepper motor. Mount the motor to the *y*-motor mount using three M3 washers and screws. (*Note*: be careful about the orientation of the pulley on the motor.)

FIGURE 5.20
Mounted coupler and the z-axis threaded rod.

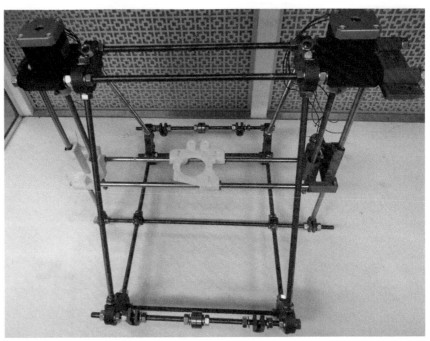

FIGURE 5.21
Released *x*-axis in context.

FIGURE 5.22
Mounted pulley on the *x*-axis stepper motor.

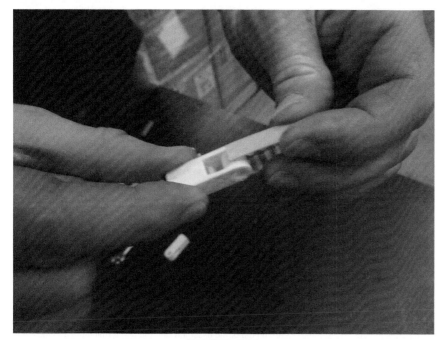

FIGURE 5.23
Attaching the belt terminator.

Assemble the BOM for the super-slim *y*-stage (shown in Figure 5.26):

Two of each carbon fiber rods[12]: 200 and 185 mm.
Four M3 × 10 mm screws.
Four M3 bolts.
Four corners, four saddles and one spacer (or print plate01.stl).
Plastic weld epoxy.
Heated bed.
Thermistor.

Next put them all together as shown in Figure 5.27, which is showing your completed goal for the stage. Start by printing or procuring the four corners, four saddles and one spacer (or print plate01.stl). Then measure carbon-fiber rods. You can cut the fiber rods with a utility knife (score the rod to length, then roll the rod under the blade while applying pressure). You can also cut it with a hacksaw, but be careful to drag it back slowly as carbon

---

[12]Carbon fiber rods are from goodwinds.com/merch/list.shtml?cat=carbon.pultrudedcarbon item no. 020961.

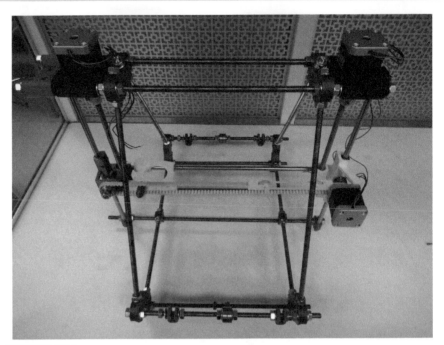

FIGURE 5.24
For proper alignment, the terminators are on top of the assembly and the belt terminators and the stage are on opposite sides of the printer.

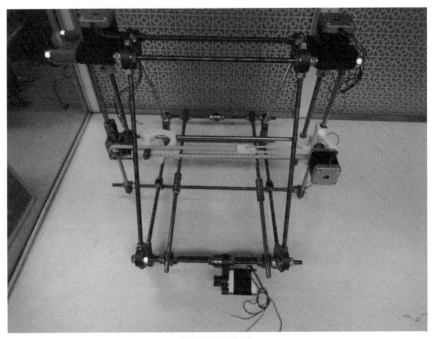

FIGURE 5.25
y-motor mount installed between the double sets of washers on the base.

FIGURE 5.26
Bill of materials for the super-slim y-stage.

fiber will tear. Next, put an M3 nut and M3 × 10 mm screw into each corner. Use the heated build platform circuit board to space the corners correctly. Then, push two saddles onto a rod and put one of the corners onto a corner of the board with a nut. Make sure to put the trace side up as the deep holes on the corners must be facing up. Next, push one end of the rod with the two saddles on it into the corner on the board. Push another corner onto the opposite end and attach it to the board. Repeat this identical process on the other side. Put the entire stage together. Make sure everything fits and is in the right direction. Make sure that the holes for the saddles are facing inward. If it does, remove one corner piece from the board and carefully apply a little epoxy to the ends of the rod, push the corner pieces back on and reattach the corner piece to the circuit board with a nut. Rotate the rod and move fore-aft to get the epoxy smeared around. Repeat the process

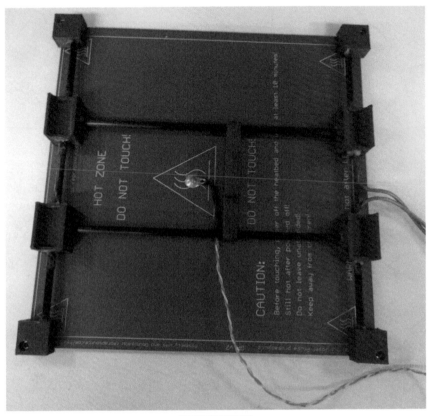

FIGURE 5.27
Assembled super-slim *y*-stage.

on the other side. Make sure to put the saddles on first before you epoxy everything!

To get to this point should take you <8 h even assuming you were taking your time and stopping for food and restroom breaks. At this point, you can either push forward on the extruder or go do something else while the epoxy dries (like sleep or if you are married, take your spouse out to dinner—trust me on this. 3-D printing is addictive particularly when you first start and it is always a good idea to keep your spouse happy.).

After the epoxy has set, move the saddles so that they are relatively equally spaced one-third of the length of the rod. Put the other two rods through the spacer. Loosen the nuts on the corner pieces, so the parallel rods can be moved slightly. Then, push the cross-rods into the holes on the saddles and tighten the nuts so that the whole assembly is relatively fixed to the board. Finalize spacing of the saddles pushing the spacer from one side to the other. When happy with their position, apply a small dab of epoxy to each saddle/parallel rod to affix the saddle to the parallel rods. Let the epoxy set. Loosen the nuts, remove the cross-rod ends from the saddles and apply epoxy to the ends of the cross-rods. Reinsert the cross-rods into the saddles and rotate them and push them side to side to smear the epoxy. Tighten the nuts, so the assembly is rigid and fixed. Again, you can skip ahead and come back to finish when it is dry or take a short break.

Turn your attention to the heated build platform. Use silver epoxy to mount a thermistor to the heating stage as shown in Figure 5.27. Space the y-axis guide rods so that the saddles fit snugly on the bearings, tighten everything up, and affix the bearings to the saddles with small zip ties. Then, align the y-axis drive belt with the dog and push the belt into the dog. Move the carriage fore and aft to assure motor and idler are aligned properly and tighten everything up. Now make sure you have everything installed correctly. Ensure that the y-axis is plumb with the x-axis. You can use calipers or measuring tape to ensure equidistance on each side between the x- and y-slide rods. (*Please note*: Never skimp on the alignment as it will become incredibly important when we get to the stage of actually printing.)

You can make the following optional design change to make leveling the bed easy in the future: if you want you can use four M3 × 20 mm screws instead of four M3 × 10 mm screws for attaching the bed. This will require eight more M3 nuts (12 total), but allows for easy bed leveling in the future. To do this, just put an M3 nut in the nut trap on the corner pieces and screw the M3 × 20 until it is tight. Then screw another M3 nut down the M3 × 20 and put the heated bed on top and secure with another M3 nut. Repeat these steps for each corner piece. You may want to do the initial setup to make sure the layup is proper and epoxy is set before changing.

Additional components you will need to complete the y-axis are the following:

> y-belt of 125 mm
> Two belt ends
> Zip tie
> M3 × 20 mm and two M3 nuts and washers

Next, it is time to mount the y-axis as shown in Figure 5.28. Orient y-motors facing you, push the nuts to the left side of the frame and tighten jam them up. Pull the excess y-axis guide rods toward the side with the motor and tighten them down. Push two wire ties through each saddle. You will find it easier to pull them through using needle-nose pliers. Mount the heated bed to your printer with the wires facing the Melzi and line up all your bearings. If the z-axis is getting in the way, move the z-axis nuts up. Next, tighten down the

FIGURE 5.28
Installed and mounted y-axis and stage.

wire ties over the bearings. Now, slide y-axis back and forth—while trying to minimize friction—and then tighten the M8 nuts in place. The y-axis should slide back and forth with very little drag—adjust the y-rods as needed and slide it back and forth until it slides smoothly.

Now you are ready to add the belts. Cut the y-belt to 125 mm in length. Trim the ends of the belt, so that it terminates with a tooth. Put belt terminators at both ends of the belt as shown in Figure 5.29 following the same procedure as the belt terminators described above. Loop the belt around the y-motor shaft and idler shaft and then tie into place with a zip tie so that it is somewhat loose.

Now, tilt the printer on its side and push the y-stage all the way to the limit switch and then pull the two belt terminators to the other side as shown in Figure 5.30. Then slip the teeth of the belt into the dog on the stage. Move the M8 nuts on the main rods to assure that the y-motor and the idler until the belt does not walk when you move the y-axis back and forth. When everything is aligned, tighten the bolts down and then pull on the zip tie to make the belt tight enough to "twang" like a guitar. Put the glass or mirror (only because it

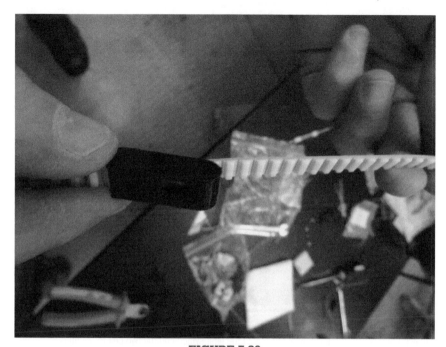

FIGURE 5.29
Placement of the belt terminators at both ends of the belt.

looks more interesting) precut to size on top of your bed and hold in place
with four small binder clips as shown in Figure 5.31.

5.2.8. The extruder
Gather the BOM as shown in Figure 5.32 for the extruder:

Wade's gear big and small.
One extruder body.
One spacer.
One extruder mount.
One hot end mount.
One motor spacer (thin).
One Bowden sheath 2 ft long (8 inch OD, eigth by a quarter ID, PTFE tubing).
Two M6 nuts.
Three 608 bearings.
Five M8 washers.
One M8 locknut.
One Hobbed Bolt.
One-inch threaded rod idler shaft.

FIGURE 5.30
Attaching the belt with the *y*-stage all the way to the limit switch and the two belt terminators to the other side.

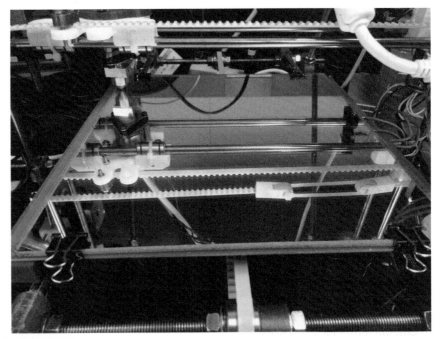

FIGURE 5.31
Installed print bed (mirror precut to size) on top of your bed and held in place with four small binder clips.

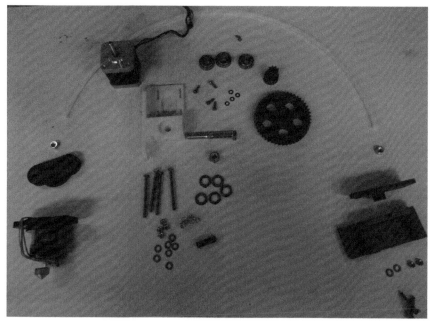

FIGURE 5.32
Bill of materials for the extruder.

Four M4 × 50 mm screws.
One Budaschnozzle 1.2 0.5 mm(LulzBot).
Four springs.
Six M4 nuts.
Ten M4 washers.
Two M4 × 20 mm screws.
One stepper motor.
Four M3 × 10 mm screws.
Four M3 washers.
One set screw.
M3 nut for small wade gear.

Use a pencil sharpener to trim down both sides of the Bowden sheath as shown in Figure 5.33. If you do not have access to a pencil sharpener, you can get the same result by whittling with a knife. Next, mount a bolt into both ends of the printed sheath holders and plastiform threads at both ends of the sheath as shown in Figure 5.34. Clean out extruder body holes and mount M6s in the holes of the extruder mount and hot end mount (Figure 5.35). Screw the long M4s screws on the extruder body and the idler carriage (Figure 5.35—put all the way on and then take off—as that piece you want to be freely moving). Now, mount the

FIGURE 5.33
Using a pencil sharpener to trim down the bowden sheath.

FIGURE 5.34
Plastiformed threads of the sheath.

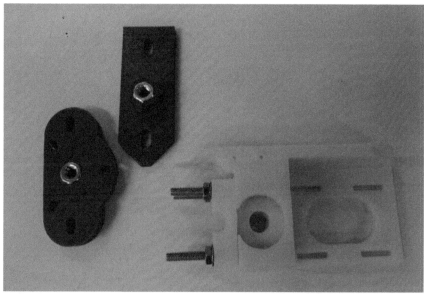

FIGURE 5.35
Mounted M6s in the holes of the extruder mount and hot end mount and the long M4s screws on the
extruder body and the idler carriage.

set screw and M3 nut in small gear and mount that on the stepper motor shaft
and mount the motor to the extruder body loosely with M3 screws as shown in
Figure 5.36. Mount the extruder body on the extruder mount before attaching it
to the frame using two M4 screws washers and nuts. This will ensure that before
you put on the gears, you will be able to mount it on the printer. *Warning*: If you
do not do this step, you will need to take the whole extruder apart.

Next, mount the main large herringbone gear by taking the hobbed bolt, gear,
2–4 washers, 608 bearing and then push through extruder body, 608 bearing,
washer and a lock nut. The number of washers will depend on your hobbed
bolt. You want your filament to be centered on the hobbed part of the bolt
(as shown in Figure 5.37). Now match up the gears (Note: This may demand
a stepper motor spacer depending on the motors you obtained.). This spacer
is shown in blue in Figure 5.38 and can be adjusted in OpenSCAD for perfect
fit. Match the gears up and screw the stepper motor to the extruder body with
two M3 screws and tighten them down. Then push the hobbed bolt out of
the large gear and catch the washers as they fall. Then screw the other two
M3 screws and washers down and put the hobbed bolt and washers back
together. Mount it all to the frame with the M8 bolts and washers.

FIGURE 5.36
Partially assembled extruder.

Now, turn your attention to the other end of the Bowden sheath. Screw the sheath down over the M4 nut in the extruder holder and trim off with knife to be flush with nut. Run a one-eighth drill bit through the hole as shown in Figure 5.39.

You may find this easier to do holding the drill bit with pliers.[13] Put the sheath through the extruder drive mount (gray) and then screw the sheath down over the M4 nut in the drive holder (blue). Trim with knife, run the drill bit, and threw it following the same procedure in Figure 5.39. Then, run the filament through the Bowden cable and make sure all PTFE shards are removed. If you skip this step, there is a risk that you will get a plugged nozzle before you even start printing. Please note again all the blue parts in the images are in ABS to provide better protection against warping at high

---

[13]If you have access to a 3-D printer already, you can also print out these handy drill bit handles. http://www.thingiverse.com/thing:91035.

FIGURE 5.37
Installed correctly, the filament is centered on the hobbed part of the bolt.

temperatures. You should now have the Bowden sheath completely constructed as seen in Figure 5.40.

The next step is to prepare the extruder. Remove the printed plastic base on
the Budaschnozzle (Figure 5.41, the base is printed in black). You may want
to keep this black piece so you can direct mount your extruder to the *x*-axis in
the future. Now attach the extruder mount (blue piece) to the Budaschnozzle
using the same screws and tighten the whole Budaschnozzle down as shown in
Figure 5.42. Attach the Budaschnozzle to the *x*-axis using two M4 × 20 screws,
nuts and washers and your printer should look like it is coming together as in
Figure 5.43.

FIGURE 5.38
The spacer used to align filament and hobbed bolt.

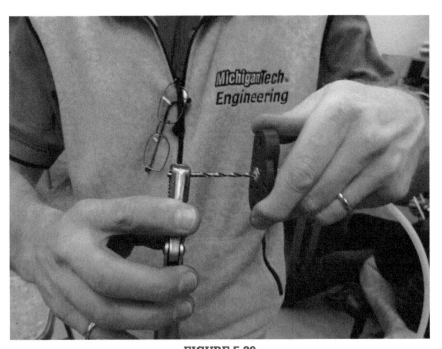

FIGURE 5.39
Preparing the Bowden sheath.

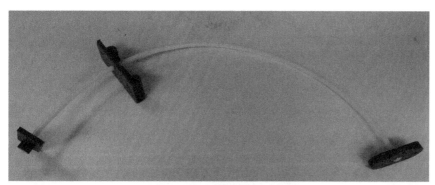

FIGURE 5.40
Completely constructed Bowden sheath.

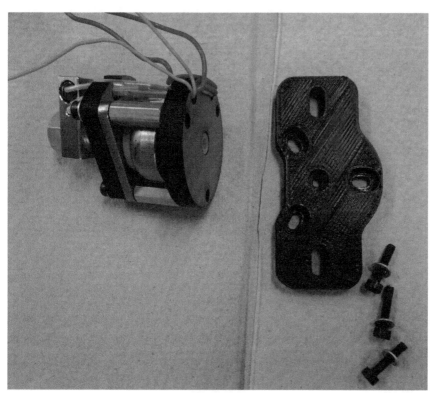

FIGURE 5.41
Printed plastic base removed from the Budaschnozzle.

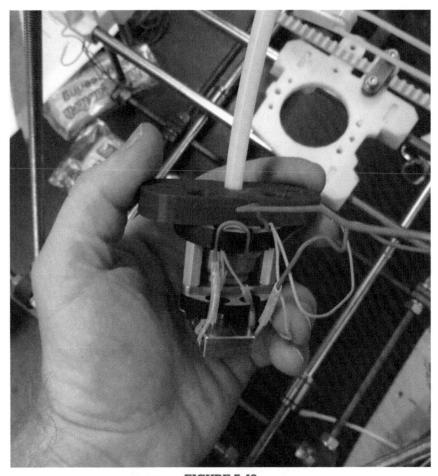

FIGURE 5.42
Extruder mount attached to the Budaschnozzle.

5.2.9.   Electronics
Assemble the BOM for the electronics as shown in Figure 5.44:

Melzi boards
Two printed Melzi mounts (printed)
Four M3 × 10 screws
Four M3 bolts
Four M3 washers
Wires
Heat-shrink tube
Lighter or heat gun
Zip ties

FIGURE 5.43
Budaschnozzle attached to the *x*-axis.

If you have never soldered before, now is your chance to learn! I recommend trying to solder two wires together first to get the hang of it. Solder the wires together tinning the surface first and do a butt joint.

When you are ready, cut the thermistor leads to have one slightly shorter than the other (a few millimeters). Get 3 ft of two conductor wires. Then cut your conductors to the same length difference. Cut 1 cm of heat shrink tubing to completely cover the first lead and a slightly larger diameter heat-shrink tube to go over the first and the other wire. String them both on the wire. Strip about 0.5 cm of wire and overlap the thermistor leads with the wire. Solder the wire to the thermistor lead by tinning the surface first—do a butt joint. Put the first heat-shrink tube down and use a lighter or heat gun to shrink it. Then solder the second wire. Finally put the longer heat-shrink tube down over the second wire and first heat-shrink tube—and shrink it in place with the lighter or heat gun.

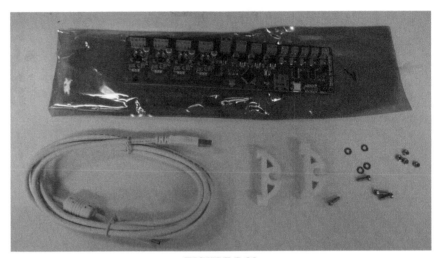

FIGURE 5.44
Bill of materials for the electronics.

Now it is time to begin building the electrical system. Screw the Melzi mounts onto the Melzi board with M3 screws, bolts and washers. Mount the Melzi board to the right of your printer on the far side of the extruder using zip ties as shown in Figure 5.45. Next, solder 3 ft of wire to the thermistor (orange) epoxied to the heated board and solder 3 ft of wire to the front metal contact on the heated board as shown in Figures 5.46 and 5.47.

Zip tie the wires to the heated board to ensure you do not stress the soldered joints as shown in Figure 5.48. Now depending on the motors you have, you may need to solder extensions to any wires that are necessary to ensure they can reach to the Melzi without running into the moving parts. Then heat shrink tube over them to clean them up or you can braid all your wires. In general, it is a good idea to have tidy wire management and spending a few extra minutes will save you from frustration in the future when you break a connection because a wire was out of place and got pulled by the printer.

When using the Melzi board, pay particular attention to polarity of the incoming power. The board is marked and black goes to ground (gnd)! Failing to wire the board properly will risk smoking (destroying) your board. In addition, motor wiring is terminal-specific. Use a multimeter to identify what motor leads belong together (pairs will have some low resistance, like $8\,\Omega$ whereas nonpairs will be essentially open circuits). Ensure that your wires are paired by using a multimeter and measure the leads on the steppers. Black and green are a pair as can be seen in the images of Figures 5.49 and 5.50.

FIGURE 5.45
Mounted Melzi board on the right of your printer on the far side of the extruder.

The rest of the wiring is not terminal-specific, so you do not need to worry about polarity. Once everything is wired up, check motor rotation with the interface described below and if it is wrong, swap conductor pairs.

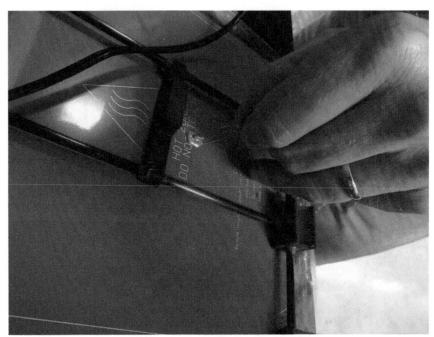

FIGURE 5.46
Soldering wire to the thermistor (orange) epoxied to the heated board.

FIGURE 5.47
Soldering wire to the front metal contact on the heated board.

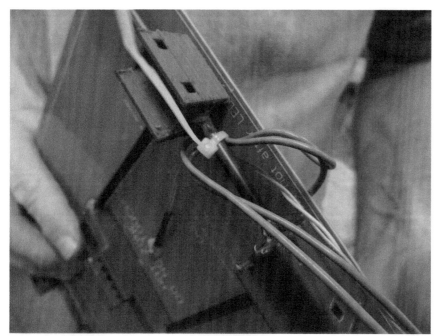

FIGURE 5.48
Zip tied wires to the heated board to relieve stress on the soldered joints.

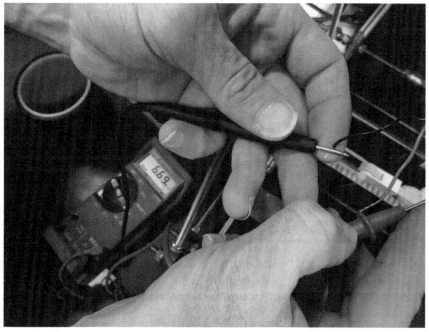

FIGURE 5.49
Unpaired wires being tested with a multimeter.

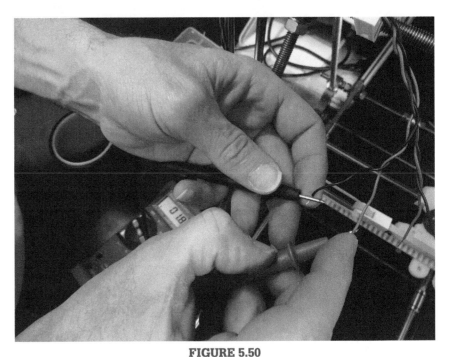

FIGURE 5.50
Paired wires being tested with a multimeter.

5.2.10. Limit switches

The next step is to gather the BOM of your limit switches as shown in Figure 5.51, which consists of the following:

Three mechanical limit switches and wires, three end stop holders (printed), three M3 × 20 mm screws and three M3 washers.
Three M3 nuts.
Six M2 × 10 mm screws and six M2 nuts.
Three M2.2 washers.

Mount your limit switches to the end stop holders. Plastiform the threads into the plastic and put on all the M2 screws. Mount the three limit switches to x-, y-, and z-axes, respectively, as shown in Figures 5.52, 5.53 and 5.54.

The x-switch is triggered by the x-carriage. The y-switch is triggered by the corner of the y-frame (blue) and the z-switch is right beneath the flexible arm (yellow). Put an M3 and two nuts on the flexible arm to enable you to do fine control of the z-switch.

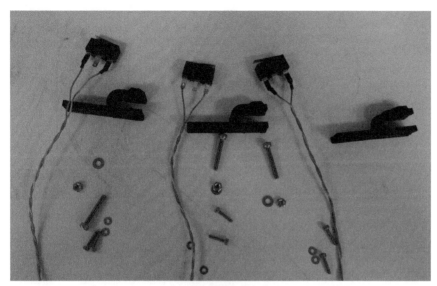

FIGURE 5.51

Bill of materials for the limit switches.

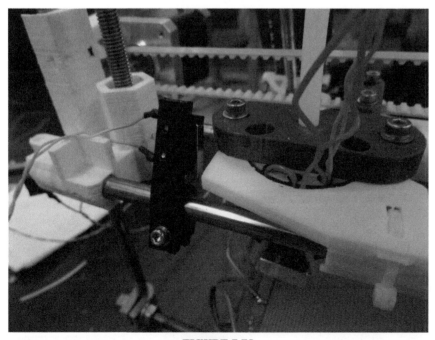

FIGURE 5.52

Mounted *x*-limit switch.

FIGURE 5.53
Mounted y-limit switch.

FIGURE 5.54
Mounted z-limit switch.

5.2.11. Power supplies

You can purchase a power supply specifically for your RepRap, but chances are if you are in either an educational institution or an industrial lab, there are a plethora of appropriate power supplies available from old computers that are being discarded. Cannibalize an old computer or buy a power supply that provides a minimum of 15 A at 12 V. After the power supply is completely unconnected, cut off the ends of all the black and yellow (12 V) supplies, group them and solder together (as seen in Figure 5.55). Take the yellow and black bundles and extend with wire by soldering. Then plug these power lines into the Melzi Board. You will not need the other leads, but leave them on to both prevent shorts and leave room for expansion in the future (e.g. adding a cooling fan).

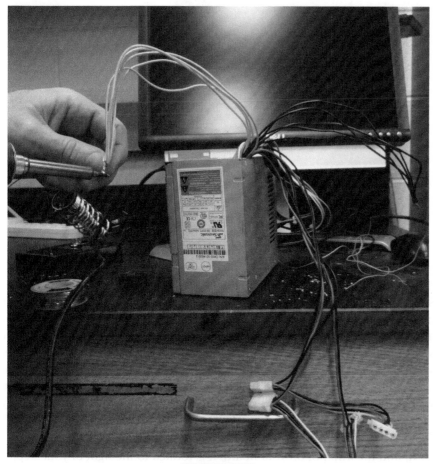

FIGURE 5.55
Grouping and soldering the power supply wires.

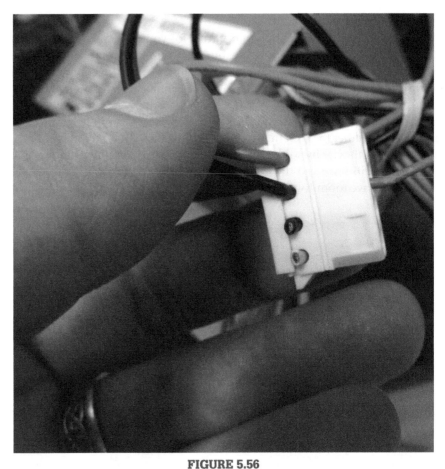

FIGURE 5.56
Shorted the green wire with black ground in order to provide the 12V from the power wires.

Short the green wire with black ground in order to provide the 12V from your black and yellow as seen in Figure 5.56. Do this by simply cutting the green wire and any black wire from the 24 pin power connecter on the power supply and soldering them together. You can then put some heat shrink tubing around the solder. If your power supply has a mechanical switch on the back, a straight solder is fine; if there is no mechanical switch (i.e. you plug it in to turn it on), you may want to wire a switch between the green and black wire to turn it on and off easily.

5.2.12.   Adjust the trim pots for motors

Next, find the trim pot on your board, it is located to the right of your stepper driver and below the large capacitors (below stepper driver and to the left of the large capacitor if you are looking at it while mounted on the frame). It will have a screwdriver like top surface. Notice that on the left side of the trim pot, there is a single large tab at the base and on the right side, there are two tabs. We are interested in the single left tab.[14] If your board is not plugged into the USB, plug it in now. You do not need to power the entire board to adjust the trim pots. Ensure that the power selection jumpers are set to USB as well. The USB power is enough to allow for trim pot voltage readings. Next, use a multi-meter and set it to DC voltage within the range for 0–1 V. Measure the voltage from the trim pot, red probe goes to the left side, and single tab of the trim pot. While the black probe goes to GRND (feel free to use the GRND from JP16). Be careful not to slip—you may short something! The voltage by factory default should be around 0.7 V or so, which is high for most motors. Melzi has a sense resistor value of 0.2 Ω, ultimately the math you need to know to set the right Vref is 0.28× (max current (A) of your motor). If you purchased the motors recommended in the BOM, it is $0.28 \times 1.7 = 0.476$. If you are really curious about the math behind this tuning, the Reprap wiki provides a very good overview.[15]

Turn the trim pot clockwise to turn down the voltage, clockwise to turn it up. *Be aware there is no stopper on the trim pots.* It is possible to over turn the trim pots, so please use a multimeter when adjusting. Read the voltage and adjust until proper Vref is set. You can also overshoot by a little bit (roughly within 0.1 V). Finally, plug in the power supply into the Melzi board and you are ready to start—black is ground.

5.3. SOFTWARE

The software tool chain consists of three discreet parts: firmware, printer interface and slicing software (although some interfaces can also provide slicing). In short, the slicing software takes a 3-D model and converts it to g-code, which is a numerically controlled programming language that tells the 3-D printer how to make your model. The printer interface sends the g-code to the printer and the firmware interprets the g-code into movements.

5.3.1. Firmware

Firmware is really a software that resides on the printer controller board (in this case the Melzi). The firmware (1) interprets g-code into motion, controlling the motors, the temperature of the hot end and heated build platform, (2) takes notice when limit switches are activated and (3) echos responses when commands are executed or a problem occurs. Ideally, you will only need to configure your firmware and upload onto the controller board once and the printer is then

---

[14]This is easiest to accomplish with two people—one measures while the other adjusts.
[15]Stepper motor driving board http://reprap.org/wiki/Pololu_stepper_driver_board.

ready for use. If you are using a 3-D printer that is known to be functional already, then it is unnecessary to complete the following steps pertaining to firmware configuration and uploading. Practically, firmware will be occasionally upgraded, so the practice of configuration and uploading is worth understanding.

5.3.2.   Printer interface

The *printer interface* is software that runs on the host computer attached to the printer via a USB cable. It provides an interface for controlling the printer: setting temperatures, moving axes, starting print jobs, etc. The printer interface essentially converts interactions with it into g-code—when a button or linked image is clicked or text is entered into the printer interface, it is converted into g-code that is then sent to the printer firmware for interpretation into some action.

5.3.3.   Slicer

A *slicer* is software that converts a solid model (typically an STL file type) into the g-code that the printer firmware can interpret into action. After slicing a model, the g-code is either loaded onto a micro-SD card, which is then inserted into the printer control board and then activated through the printer interface or the g-code is loaded into the printer interface and then transmitted to the printer controller board in small chunks over via the USB connection.

5.3.4.   Software sources

There are five software packages you will need to download (three for the firmware and one each for the printer interface and slicing). All the software is free and open source:

1. Firmware: Marlin.[16]
2. Arduino IDE.[17]
3. Melzi driver[18] and instructions.[19]
4. Printer interface: Pronterface.[20]
5. Slicing software: Cura.[21]

The primary repository for 3-D printable designs, as of this writing, is Thingiverse,[22] although it should be noted there is an ongoing discussion for making

---

[16]Marlin https://github.com/ErikZalm/Marlin/tree/Marlin_v1/Marlin.
[17]Arduino IDE http://arduino.cc/en/Main/Software.
[18]Melzi board https://github.com/maniacbug/mighty-1284p.
[19]Melzi instructions http://matterfy.com/pages/how-tos.
[20]Printrun https://github.com/kliment/Printrun.
[21]Cura http://daid.github.com/Cura/.
[22]Thingiverse http://www.thingiverse.com/.

a more appropriate central repository of enabling innovation.[23] I keep two collections of designs updated that will be of most interest to the readers of this book:

1. Open-source scientific tools.[24]
2. Open-source optics.[25]

Both collections are growing rapidly. You can also search Thingiverse by keyword, or by browsing click Explore, Categories, Learning. Unfortunately, as of this writing, Thingiverse only has learning broken down into the following categories: biology, engineering, math, physics and astronomy. If there is something that you want to print that is not up yet, please design it yourself and post it!

5.3.5. Controller board installation

Although, I would strongly recommend that you move your computer to an open-source operating system like Linux (such as Debian[26]or Linux Mint[27]) and you can also use your RepRap with a Windows or Apple computer. If you are not using Linux and a package installer to install Arduino easily, you can download the Arduino integrated development environment (IDE) for your operating system[28] and follow their *Getting Started* guide for your operating system. Once you are able to open the Arduino IDE, download the Melzi driver from the above link and follow instructions to save the driver in the Arduino folder. Upon restarting the Arduino IDE, the Mighty 1284P 16 MHz using Optiboot should be an option under Tools→Board. Select it.

For a Linux installation, plug the Melzi board into an available USB port on your computer using the supplied cable. Install the arduino and arduino-mighty-1284p packages using your package manager. Upon starting the Arduino IDE, the Mighty 1284P 16MHz using Optiboot should be an option under Tools→Board. Select it.[29]

For the windows driver installation, plug the Melzi board into an available USB port on your computer using the supplied cable. If prompted to install a driver,

---

[23]Improved digital design database http://www.thingiverse.com/thing:30880.

[24]Collection of open-source scientific tools that re 3-D printable [16] http://www.thingiverse.com/jpearce/collections/open-source-scientific-tools.

[25]Collection of open-source optics tools that are 3-D printable [33] http://www.thingiverse.com/jpearce/collections/open-source-optics.

[26]Debian http://www.debian.org/.

[27]Linux Mint http://www.linuxmint.com/.

[28]Arduino softwarehttp://arduino.cc/en/main/software.

[29]There may be a warning that you should be a member of the dialout group. If so, follow the instructions.

select manual installation and locate the driver in Arduino folder under drivers under "FTDI USB drivers"

For an Applie installation, follow all operating system instructions, no extra driver needed.

Again on any operating system other than Debian, you will need to ensure mighty-1284p installed, so Arduino will understand the board.

5.3.6.   Firmware configuration

The firmware is written for Arduino and Arduino-based controllers in a programming language known as C++ (as discussed in detail in Chapter 4). The firmware used for the MOST RepRap builds is Marlin, available at the link above. To install and configure, do the following:

1. Download Marlin firmware from the above link and unzip into a logical and memorable location. As in Chapter 4, it is easier if this location is the sketchbook folder.
2. Open the Arduino IDE and then open Marlin.pde from the File→Open… option, navigating to the location where Marlin was unzipped to.
3. Note that the IDE opens a number of different tabs, select the tab Configuration.h.
4. Locate the line #define BAUDRATE and replace number with 115200.
5. Scroll down the page and locate #define MOTHERBOARD and replace the number with 63 to set the firmware for the Melzi board.
6. Scroll further down and locate the line #define TEMP_SENSOR_0 and set it equal to 1.
7. Set #define TEMP_SENSOR_BED = 4.
8. Locate the line starting with #define X_ENDSTOPS_INVERTING, set it and the two following (Y and Z) to false.
9. Locate the line starting #define INVERT_X_DIR. Follow the instructions in the remarks for the six lines starting at that entry for Mendel printers and geared extruder.
10. Scroll further down locating the line #define DEFAULT_AXIS_STEPS_ PER_UNIT. The values in the parentheses represent an array; a number for each axis is separated by a comma. These settings tell the printer how many steps are required to move a millimeter of whatever the axis moves (x, y and z move part or hot end, E moves filament). Using the RepRap calculator from Josef Prusa,[30] we can figure out how many steps are required for x, y and z. The motors are 0.9°/step (although

---

most users use 1.8°/step motors as they work better and can handle more torque) and the Melzi board is set to use 1/16 microsteps, yielding 106.667 for the 12 tooth pulleys on the x- and y-axes and 5120 for the z-axis. Trial and error has yielded a setting of 1500 for E. For 400 steps/revolution motors, this entry should be (106.667, 106.667, 5120, 1500); for 200 steps/revolution motors, it should be (53.333, 53.333, 2160, 750). *Only if using* 400 steps/revolution motor: locate the line #define DEFAULT_MAX_FEEDRATE and set the third number in the list to 2.

11. Save the updates (File→Save, ctrl-S or click the down arrow in the tool bar) and upload to the board (File→Upload, ctrl-U or click the right arrow in the tool bar). Watch the prompt cues at the bottom of the window and wait for the message "Done Uploading".

12. Once the firmware is uploaded to the controller board, open the serial monitor (Tools→Serial Monitor or the magnifying glass on the tool bar). Set the communication rate to 115200 and click the reset button on the Melzi controller board. The current printer settings should be echoed on the serial monitor screen.

5.3.7. Printer interface and slicer

The Melzi board uses a printer interface to send print jobs to the controller. Download and install Pronterface from the above link (or simply install it with the package manager if using Linux). Note that there are precompiled binaries for different OSs if you scroll down the page; if you are not using Linux, it is easiest to simply download the appropriate one for your operating system. Once Pronterface is running, select the communication port, set the speed to 115200 and click the connect button. A message will be echoed by the controller indicating that the printer is online if the connection is successful. You can test this by trying to move the printer head in the x- or y-direction.

Three-dimensional solid models are sliced into individual layers for the printer by a slicing program (slicer). Download and install Cura from the link above. A screenshot of Cura is shown in Figure 5.57 and it is relatively straightforward to use. There are also other slicing programs available and you may want to try them out to see what works best for your group.[31]

---

[31]For example, Slic3r, Skeinforge, Repsnapper, RepRapPro Slicer, etc. For direct links, see http://reprap.org/wiki/Useful_Software_Packages#Software_for_dealing_with_STL_files.

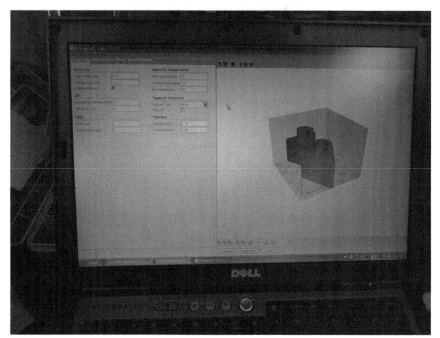

FIGURE 5.57
Screen shot of Cura.

5.4. PRINTING FOR THE FIRST TIME

You are now ready to print for the first time. Load the filament through the tube while the hot end is still cold. Ensure that your motor leads are all wired correctly by (1) opening Pronterface, (2) connecting to the printer and using the x, y, and z controls to test directionality. Do this in small steps—particularly for the z-axis. If any of the motors are spinning the wrong direction, turn off the power so you do not destroy the stepper driver, and then switch the leads.

5.4.1. Level the z-axis

Position the hot end to the far right of the board and move the nozzle down until it is two paper widths from the stage. This is the home position. You can adjust this position to perfection using a combination of moving the z-end stop, the z-micropositioner (Figure 5.58), and the bolts holding up the z-axis.

Then move the hot end to the left of the stage and adjust the bolts on the left z-axis to make it level with the right. The hot end should go from the left to right and back the exact same height from the bed.

FIGURE 5.58
The z-micropositioner.

5.4.2.   Print fantastic and inexpensive scientific hardware

You are now ready to print. You can download your first designs from Thingiverse or another repository (e.g. grabcad, github.com, 3Dhacker.com, appropedia.org, etc.). The next chapter has many examples, but first consider the simple open-source lab jack[32] shown in Figure 5.59.

This lab jack was designed by MOST after we had received an absurd $1000 quote for one to help align an analytical instrument to test solar photovoltaic devices. Lab jacks come in highly variable precision and quality as there are high-precision jacks available in the hundreds (or even a thousand!) all the way down to $25 for poorly built toy lab jacks on the Internet. As you can see in Figure 5.59, the labjack is constructed from a few relatively easy prints in PLA, ABS or other thermoplastic (e.g. no overhangs, advanced materials or complicated multihead printing is necessary). The remainder of the BOM is common screws and bolts that you are likely to have on hand having just built

FIGURE 5.59
Open-source lab jack.

---

[32]Open-source labjack http://www.thingiverse.com/thing:28298.

a RepRap and ordered nuts and bolts by the bag. The design shown here is highly customizable and perhaps offers better value than even some of the high-end lab jacks as you can control the precision and the frames by altering the design. Thus you can customize the lab jack for your specific application and build in holders, clips, bolt holes, etc. to match the specific requirements of your experiments.

The RepRap can print a lot more than a lab jack and currently there are hundreds of components of scientific instruments proliferating on the Internet. The MOST RepRap described above can be built by two people in a day for under $600. As the investment in a RepRap is capable of repaying itself after a few hours of printing even common scientific hardware, it is likely to become an indispensable resource for most laboratories as more sophisticated and customized open-source scientific tools are developed. In Chapter 6, we will look at many examples of open-source scientific hardware using RepRap printable components (or entire tools) in physics, engineering, environmental science, biology, and chemistry. This list is expanding rapidly and is meant to provide a snapshot of the current abilities to inspire you to join the open-source scientific hardware community and create your own equipment for the benefit of all of science.

REFERENCES

[1] Upcraft S, Fletcher R. Assem Autom 2003;23(4).

[2] Gibson I, Rosen DW, Stucker B. Phys Procedia 2010;5.

[3] Petrovic V, Gonzalez JVH, Ferrando OJ, Gordillo JD, Puchades JRB, Griñan LP. Additive layered manufacturing: sectors of industrial application shown through case studies. Int J Prod Res 2011;49(4):1061–79.

[4] Gebhardt A, Schmidt F, Hötter J, Sokalla W, Sokalla P. Phys Procedia 2010;5(2).

[5] Crane NB, Tuckerman J, Nielson GN. Self-assembly in additive manufacturing: opportunities and obstacles. Rapid Prototyping J 2011;17(3):211–7.

[6] Economist 21 April, 2012 [special issue].

[7] Jones R, Haufe P, Sells E. RepRap - the replicating rapid prototyper. Robotica 2011;29(1):177–91.

[8] Gershenfeld N. Fab: the coming revolution on your desktop – from personal computers to personal fabrication. New York, NY: Basic Books; 2005.

[9] Corney J. The next and last industrial revolution? Assem Autom 2005;25(4):257.

[10] Malone E, Lipson H. Fab@Home: the personal desktop fabricator kit. Rapid Prototyping J 2007;13(4):245–55.

[11] Pearce JM, Blair CM, Laciak KJ, Andrews R, Nosrat A, Zelenika-Zovko I. 3-D printing of open source appropriate technologies for self-directed sustainable development. J Sustainable Dev 2010;3(4):17–29.

[12] Bradshaw S, Bowyer A, Haufe P. The intellectual property implications of low-cost 3D printing. ScriptEd 2010;7(1):1–27, doi.org/10.2966/scrip.070110.5.

[13] Holland D, O'Donnell G, Bennett G. Open design and the RepRap project. 27th International Manufacturing Conference. ; 2010. p. 97–106.

[14] Weinberg M. It will be awesome if they don't screw it up. 2010. http://nlc1.nlc.state.ne.us/ epubs/creativecommons/3DPrintingPaperPublicKnowledge.pdf.

[15] Cano J. The Cambrian explosion of popular 3D printing. Int J Artif Intell Interact Multimedia 2011;1(4):30–2.

[16] Pearce Joshua M. Building research equipment with free, open-source hardware. Science 2012;337(6100):1303–4.

[17] Hodgson G. Interview with Adrian Bowyer. RepRap Magazine 2013(Issue 1):8–14.

[18] Bowyer Adrian. Wealth without money. ; 2004; RepRap Wiki. Available from: http://reprap.org/wiki/Wealth_Without_Money.

[19] RepRap wiki User: Emmanuel, The RepRap Family Tree - Timeline 2006–2012 v4.0 14-04-2012, Available from: http://www.reprap.org/wiki/RepRap_Family_Tree.

[20] Sells E, Bailard S, Smith Z, Bowyer A. RepRap: the replicating rapid prototype maximizing customizability by breeding the means of production. Cambridge, MA: MCPC 2007; 2007, No. MCPC-045-2007.

[21] Arnott R. The RepRap project – open source meets 3D printing. Computer and information science seminar series. ; 2008.

[22] Kentzer J, Koch B, Thiim M, Jones RW, Villumsen E. An open source hardware-based mechatronics project: the replicating rapid 3-D printer. 2011 4th International Conference on Mechatronics. ; 2011. p. 1–8.

[23] Arduino homepage; 2013. Available from: http://www.arduino.cc/.

[24] Baechler Christian, DeVuono Matthew, Pearce Joshua M. Distributed recycling of waste polymer into RepRap feedstock. Rapid Prototyping J 2013;19(2):118–25.

[25] Gonzalez-Gomez J, Valero-Gomez A, Prieto-Moreno A, Abderrahim M. A new open source 3D-printable mobile robotic platform for education. Adv Autonomous Mini Robots 2012:49–62.

[26] Anzalone GC, Glover AG, Pearce JM. Open-source colorimeter. Sensors 2013;13(4):5338–46.

[27] Leigh SJ, Bradley RJ, Purssell CP, Billson DR, Hutchins DA. A simple, low-cost conductive composite material for 3D printing of electronic sensors. PLoS One 2012;7(11):e49365.

[28] Redlich TOBIAS, Wulfsberg JP, Bruhns FL. Virtual factory for customized open production. Tagungsband 15th International Product Development Management Conference. 2008; Hamburg.

[29] DeVor RE, Kapoor SG, Cao J, Ehmann KF. Transforming the landscape of manufacturing: distributed manufacturing based on desktop manufacturing (DM)2. ASME DC J Electron Packag July 18, 2012;134(4)http://dx.doi.org/10.1115/1.4006095.

[30] Moses Matt, Yamaguchi Hiroshi, Chirikjian Gregory S. Towards cyclic fabrication systems for modular robotics and rapid manufacturing. Proc Rob: Sci Syst 2009.

[31] Bayless Jacob, Chen Mo, Dai Bing. Wire embedding 3D printer. ; 2010.

[32] Bhatia Sangeeta N, Chen Christopher S. Rapid casting of patterned vascular networks for perfusable engineered three-dimensional tissues. 2012.

[33] Zhang C, Anzalone NC, Faria RP, Pearce JM. Open-source 3D-printable optics equipment. PLoS ONE 2013;8(3):e59840.

CHAPTER 6

Digital Designs and Scientific Hardware

6.1. OPENSCAD, REPRAP AND ARDUINO MICROCONTROLLERS

This chapter provides examples of scientific equipments produced using the open-source hardware (OSH) paradigm. This section will provide the basic background in the use of the three primary toolsets for OSH equipment. First, the use of parametric open-source designs using an open-source computer-aided design (CAD) package is described to customize scientific hardware for any application. This tool will then be used as the example for many scientific and engineering disciplines throughout the chapter whenever possible. Second, details are provided on how to use the open-source 3-D printers discussed in Chapter 5 to fabricate both the primary components of scientific tools and how to construct complex, multicomponent engineering and scientific devices. Finally, open-source electronic prototyping platforms, such as were discussed in Chapter 4, will be demonstrated to control complex scientific devices saving your lab time, money, and improving your research productivity.

6.1.1. OpenSCAD

Although the RepRap can print STereoLithography (STL) files generated from any CAD package, the majority of the physical designs in this chapter were developed in the free and open-source OpenSCAD[1] as it is currently the defacto "maker" standard.[2] It is the most flexible currently available

---

[1]OpenSCAD http://www.openscad.org.

[2]There are many other free and open-source CAD tools available such as BRL-CAD, Blender CAD, Python CAD, OpenCASCADE, FreeCAD, HeeksCAD, Wings3D, and Shapesmith. In addition, there are some open-source drawing packages like Art of Illusion and Blender, which are useful. Finally, there are free but closed-source tools like Google Sketchup and TinkerCAD.

As of this writing, none of the open-source CAD packages are as advanced as the commercial CAD software that, for example, mechanical engineering students are accustomed to using (e.g. Solid Works, NX, AutoCAD, Inventor, etc.). If you have a student researcher already familiar with these tools, it may save your group time in the short term to do your designs in closed software and if you still post the STLs, they may still be somewhat useful to the open-source community. However, to take complete advantage of the open-source paradigm, which will build upon and improve your designs for your benefit, the tools that are used to create the designs also need to be open so you are including the greatest possible number of collaborators. It is also interesting to note as 3-D printing expands the number of software developers assisting on the open-source CAD packages will expand, and if history is a basis of what will happen in the future, the open-source packages will surpass what is available both now and in the future in closed form.

Open-Source Lab. http://dx.doi.org/10.1016/B978-0-12-410462-4.00006-8

open-source CAD package and you do not necessarily need to know anything about conventional CAD. OpenSCAD is an open-source, script-based CAD application. It is used by writing simplistic code to describe the geometric specifications of the required object by using three primitive shapes (cylinder, sphere and cube) and complex polygons using the polygon, polyline and the 2-D and 3-D extrusion commands. If you understand the basic geometry and Boolean logic, you should not have any trouble getting started. If you have written any kind of computer code before, then this will be easy. The real power of OpenSCAD is that it allows for parametric designs, which is the ability to alter a design to specifications by changing the parameters of the geometry of an object. This allows changes to be made to the design easily and quickly by simply changing the value of user-defined variables. Simply put—if you change a single number in the code—you can immediately have a new customized design. This would be great enough on its own, but Thingiverse.com, the repository that now holds well over 100,000 OSH designs, has a built in Customizer App that enables you make it absurdly easy for others to customize your designs if they are written in OpenSCAD. The would-be custom users of your design do not even need to go into OpenSCAD to alter the design, they can make the change from their browser window. To illustrate briefly how OpenSCAD works, consider the code for a simple gasket[3] below:

```
//Customizable Gasket
// This is a simple open-source parametric gasket
//CUSTOMIZER VARIABLES
//Defines the height of the gasket
gasket_thickness=1; //[1:10]
//Defines the gasket outer diameter
gasket_outer_diameter=2; //[2:100]
//Defines the gasket inner diameter
gasket_inner_diameter=1; //Numeric value smaller than Gasket Outer
Diameter
//CUSTOMIZER VARIABLES END
module gasket()
{
  difference()
  {
  cylinder(h=gasket_thickness, r=gasket_outer_diameter*2,
  center=true, $fn=100);
  cylinder(h=gasket_thickness, r=gasket_inner_diameter*2, cen-
  ter=true, $fn=100);
  }
}
gasket();
```

---

[3]Customizable gasket http://www.thingiverse.com/thing:58665.

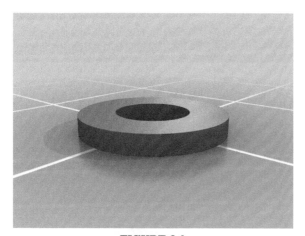

FIGURE 6.1
A 3-D rendering of a simple parametric gasket.

This code produces a simple gasket as shown in Figure 6.1.

In this code, anything following a line that begins with // is commented out and ignored when generating the STL image. The gasket is developed by making two simple shapes (cylinders) and subtracting one from the other using the difference command. For a complete tutorial on how to use OpenSCAD, see the OpenSCAD User Manual,[4] which is an open wiki book and thus constantly updated and improved. Although making a simple gasket could be accomplished in less lines of code and comments, this code is written to be as useful for others as possible. First, it is well documented explaining what every variable is. In addition, it is put in the format to be able to be used for the Thingiverse customizer.[5] When the code is opened in the customizer, it looks like Figure 6.2.

Finally, the code is set up as a module so others can copy it and put it directly into a more complex design. This way anyone wishing to make a custom gasket can adjust the size in the interface and does not even need to go down into the code. This may seem like overkill for this simple design, but as the code becomes more complex, this is a really nice favor to do for those who may need it in the future (may be even yourself).

To illustrate the utility of parametric design for a scientist, consider an optical chopper wheel, which modulates the frequency of a light beam. For a given

---

[4]OpenSCAD User Manual http://en.wikibooks.org/wiki/OpenSCAD_User_Manual.
[5]For more information about coding for the Customizer, see http://www.makerbot.com/ blog/2013/01/23/openscad-design-tips-how-to-make-a-customizable-thing/.

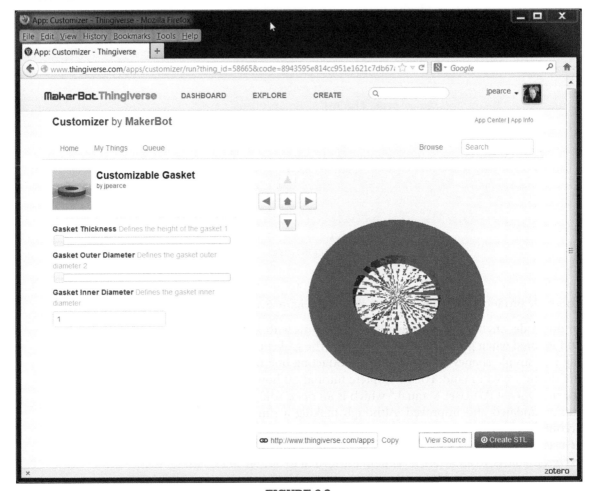

FIGURE 6.2

Screenshot of the parametric gasket code is the Thingiverse Customizer App.

chopper wheel setup, the target frequency can be adjusted by the number of slots in the chopper wheel. Normally, this involves the separate purchase of a new wheel for each experiment, and the experimenter does not always know beforehand what the optimal operational frequency is for a given application. A parametric design from OpenSCAD is shown in Figure 6.3, where the slot number for an open-source optical chopper wheel[6] has been adjusted by changing a single variable in the code from (a) 10, (b) 15 and (c) 60 slots providing ranges of chopper frequency of 20 Hz–1 kHz, 30 Hz–1.5 kHz, and 120 Hz–6 kHz,

---

[6]Optical Chopper Wheel http://www.thingiverse.com/thing:28121.

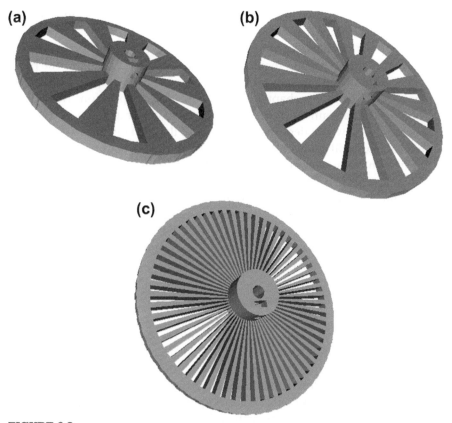

FIGURE 6.3
Rendered parametric design in OpenSCAD on an open-source optical chopper wheel with (a) 10 slots, (b) 15 slots and (c) 60 slots.

respectively. OpenSCAD directly exports the geometry to an STL file, which is used for 3-D printer open-source slicing programs (e.g. Skeinforge,[7] Cura[8] or Slic3r[9] as discussed in the last chapter), which is in turn transformed to g-code, which provides the vectors for tool path (3-D printer extruder head).

6.1.2. Open-source 3-D printers

The development of open-source 3-D printers like the RepRap[10] [1], the streamlined design of which as shown in Chapter 5 can be constructed for <$600,

---

[7]Skeinforge http://fabmetheus.crsndoo.com/wiki/index.php/Skeinforge.
[8]Cura http://daid.github.com/Cura/.
[9]Slic3r http://slic3r.org.
[10]RepRap http://reprap.org.

have made the cost of rapid prototyping accessible to most university laboratories [2]. RepRap's open-source and self-replicating nature (approximately 50% of its own parts can be self-printed) makes it an extremely useful platform for open-source fabrication and maintenance of laboratory equipment. The printing process for the additive layer manufacture of scientific experimental components discussed in this chapter is a sequential layer deposition. The RepRap extruder intakes a filament of the working material,[11] heats it, and extrudes it through a nozzle where deposits a 2-D layer of the working material, then the Z (vertical) axis will raise, and the extruder will deposit another layer on top of the first. In this way, it can build three-dimensional models from a series of two-dimensional layers [3]. Figure 6.4(a) shows the detail of a RepRap printing out a component of a filter wheel system,[12] which will be discussed in more detail below. Figure 6.4(b) shows another component of the same filter wheel printing, and also displays the assembly in OpenSCAD on a computer running only open-source Linux software. These models can be customized for a given optical setup using OpenSCAD and printed.

6.1.3.   Open-source microcontrollers

The majority of the current versions of the RepRap are controlled by an Arduino microcontroller board (or a spin-off microcontroller board), which is prototyping platform based on the ATMEL ATmega328, a high-performance, low-power AVR 8-Bit microcontroller. Arduino microcontroller boards[13] are a family of open-source, low-cost integrated circuits that contain a core processor, memory and analog and digital input/output (I/O) peripherals. As shown in Chapter 4, Arduino microcontroller boards are relatively easy to use. Thus, they have been applied in a vast number of science and engineering areas including, for example, optical tools such as holographic microscope [4], portable system for high-speed multispectral optical imaging [5], and an light-emitting diode (LED) stimulator system for vision research [6]. Here the automation capabilities of the Arduino can be combined with the rapid prototyping of the RepRap to make sophisticated, customized equipment such as an automated filter wheel changer printing shown in Figure 6.4(a) and (b).

6.2.   PHYSICS: OPEN-SOURCE OPTICS

This section introduces a library of open-source 3-D-printable optics equipment, which can be used as a flexible, low-cost public-domain tool set for

---

[11]Acrylonitrile-butadiene-styrene (ABS) and polylactic acid (PLA) polymers are the two most readily available and mature printing filament materials although many others are available and under development as discussed in the last chapter.

[12]Open-source automated filter wheel http://www.thingiverse.com/thing:26553.

[13]Arduino http://www.arduino.cc.

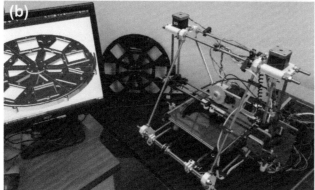

FIGURE 6.4

An open-source self-replicating rapid prototyper (a) printing a 3-D optical component—filter bracket and (b) printing a pie-slice of the filter wheel with the entire assembly displayed in OpenSCAD on a computer running the open-source Linux operating system.

developing both research and physics teaching optics hardware [7]. Overall, this section describes open-source optics development, system requirements, features, advantages, and known limitations. This same model can be applied to other physics-related research tools and equipment.

The open-source optics library is built from standard low-cost parts available in most hardware stores and customizable printed parts. First, an optical rail is fabricated from an open-source aluminum extrusion system called Open-Beam[14] as shown in Figure 6.5.

---

[14]OpenBeam http://openbeamusa.com/.

FIGURE 6.5
The cross-section and side wall of the OpenBeam aluminum extrusion system.

FIGURE 6.6
Open-source optical rail fabricated from OpenBeam using a printed magnetic base.

An optical rail is a long, straight, sturdy rail onto which optical components such as light sources and lenses can be bolted down and easily shifted along the length of the rail. Commercial optical rail sells for around $380/m ($115/ft), while Open-Beam is available from $10–12/m. The OpenBeam system is an open-sourced, miniaturized t-slot construction system that utilizes standard metric (M3) nuts and bolts to connect to an extruded 1 m aluminum rod (M3s can also be screwed into the center hole at the end).[15] The OpenBeam is converted into an optical rail

---

[15]For those of you that just built a RepRap following the instructions in Chapter 5, you probably have a lot of M3-sized components left over.

FIGURE 6.7

Open-source optical rail fabricated from OpenBeam using printed T-brackets.

using a printable magnetic base[16] (Figure 6.6), which holds the beam securely on a steel table or desk. For those not using metal tables, two OpenBeam T-brackets can either be purchased or printed to mount to any nonmetallic surface[17] (Figure 6.7).

Optical components are attached via 8-mm-diameter smooth rods centered directly above the beam using a simple rod holder[18] (Figure 6.8) or via an offset rod holder[19] (Figure 6.9) depending on your experiment. In some experimental setups, a vibration-isolated optical table is a necessity; however, for many experiments, this is costly overengineering. While avoiding the substantial cost of an optical table, some optical experimental setups with a nonlinear experimental optical setup can be fabricated using a magnetic optics base[20] as seen in Figure 6.10, which has a small cylinder opening in the bottom meant to glue in a magnet. If your RepRap is highly tuned, you can actually use a simple press fit to eliminate the need for any glue. The optics base quasi-permanently holds the position of the optical component on a magnetic surface with more flexibility than a standard optical table although not as much stability. It should be pointed out again here that all the components shown are parametric in the OpenSCAD design. So, for example, other researchers can easily alter the

[16]Open-source Optical Rail from OpenBeam—Magnetic Base http://www.thingiverse.com/thing:30729.

[17]OpenBeam T Bracket http://www.thingiverse.com/thing:30524.

[18]OpenBeam Optical Rail Simple Rod Holder http://www.thingiverse.com/thing:31370.

[19]Open-source Optical Rail Mount for OpenBeam http://www.thingiverse.com/thing:30491.

[20]Open-source magnetic optics base http://www.thingiverse.com/thing:28133.

FIGURE 6.8
Printed simple rod holder for the open-source optical rail fabricated from OpenBeam.

FIGURE 6.9
Printed off-set rod holder for the open-source optical rail fabricated from OpenBeam.

design shown in Figure 6.10 to hold a smaller or larger magnet and a smaller or larger smooth rod.

Numerous components can then be coupled to the 8-mm-diameter smooth rods and adjusted in the z-vertical axis to build up an optical assembly including

FIGURE 6.10
Magnetic optics base with rod holder.

a static filter or lens square[21] (Figure 6.11) or circular holder[22] (Figure 6.12), kinematic mirror or lens holder[23] (Figure 6.13), static fiber-optic holder[24] (Figure 6.14), screen holder[25] (Figure 6.15) and sample holder[26] (Figure 6.16). The filters in the square holders are fixed in position using M3 screws and nuts and a filter bracket, while for the circular holder, a set screw is applied as shown in Figure 6.12.

Both holders can be adjusted for other sizes of filters or lens easily using Open-SCAD. The kinematic mirror or lens mount shown in Figure 6.13 was developed by Thingiverse user ordaos and is partially parametric design used for

---

[21]Square Filter Holder for Open-source Optics http://www.thingiverse.com/thing:31483.

[22]Open-source round lens holder http://www.thingiverse.com/thing:26752.

[23]Kinematic mirror/lens mount http://www.thingiverse.com/thing:30727.

[24]Open-source fiber optic holder http://www.thingiverse.com/thing:28187.

[25]Screen holder for OpenBeam optical rail http://www.thingiverse.com/thing:31403.

[26]Simple semiconductor sample holder v3 http://www.thingiverse.com/thing:30688.

FIGURE 6.11
Static square filter holder.

FIGURE 6.12
Static circular filter holder.

FIGURE 6.13
Kinematic mirror or lens holder.

steering optics. It contains a living hinge in one corner and two magnets in the adjacent corners, which are attracted to a pair of set screws used to adjust the mirror angle. The static fiber-optic holder was designed to hold a 7.7-mm-diameter fiber-optic cable. The screen holder is designed to support a screen or card and can be mounted on optical rail with an M3 screw–nut pair. This semiconductor sample holder shown in Figure 6.16 is designed to hold a semiconductor wafer piece on a smooth 8-mm-diameter rod (again leftover parts from RepRap builds covered in the last chapter). It allows you to change samples easily using tweezers with only one hand. More complex, multicomponent optics equipment setups can also be fabricated using this method such as an open-source lab jack,[27] which is a height-adjustable platform for mounting optomechanical subassemblies as seen in Figure 6.17. Again, you can easily adjust or customize the platforms of the lab jack to hold any other type of mount for your own experiments. This makes the open-source lab jack far superior in terms of customized utility to the standard ones that in general are just simple flat platforms.

---

[27]Open-source lab jack http://www.thingiverse.com/thing:28298.

FIGURE 6.14
Static fiber-optic holder.

3-D printing and the Arduino can be combined to create automated dynamic optics systems such as the open-source parametric filter wheel[28] shown in Figure 6.18. This version of the device has a printable wheel with eight filter slots equally spaced by 45°, an Arduino Uno microcontroller, a stepper motor, a discrete optical switch flag, a sensor composed of an output, two ends and a light beam that goes from an end to the other. When the light beam is blocked, the output signal changes depending on the logic used (e.g. from high to low). Through a computer interface, the user can set the wheel position to the desired filter by clicking on the buttons displayed on the screen, also an indicator provides the current position of the filter wheel. The user's input is read by the Arduino's serial communication port, the optical switch flag denotes the filter

---

[28]Automated open-source automated filter wheel http://www.thingiverse.com/thing:26553.

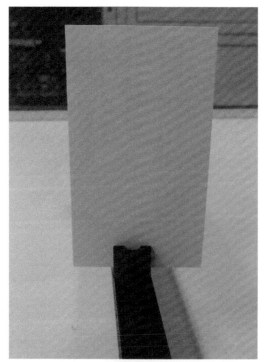

FIGURE 6.15
Screen holder.

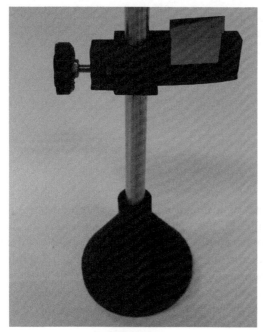

FIGURE 6.16
Sample holder.

FIGURE 6.17
Open-source lab jack.

FIGURE 6.18
Parametric automated filter wheel changer.

wheel's origin (e.g. assuming the filters are indexed from 0 to 7, the filter 0 is detected when it passes through the sensor). Each time the device is turned on or restarted, it rotates until it finds the origin. Using the current location of the filter stored inside the memory and the optical switch flag information as feedback, the Arduino interprets the input and drives the stepper motor following the logic created by the program code until the desired filter is positioned and the input matches the output. The logic always uses the shorter path until the

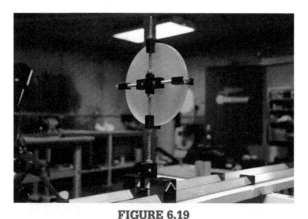

FIGURE 6.19

3-D printed optical mounts for extruded aluminum angle.

next filter by calculating the difference between the next and the current filter, and always passes through the origin when it is possible to maintain calibration. This automated filter wheel, which only costs about $50 to make, saved my laboratory from spending $2500 + shipping. It was exactly what my lab needed and having made the initial investment in the design, we will never need to purchase one again.

6.2.1. Benefits of the open-source approach for physics research

The OpenSCAD designs and STL files for the entire open-source optics library were posted on Thingiverse (see footnotes for direct links above), a digital design repository, which also provides links to all the control software for the Arduino. In addition, I maintain a Thingiverse collection of open-source optics equipment that is routinely updated as researchers from all over the world share their own designs.[29] For example, Thingiverse user kovo recently shared an even lower cost optical rail system, which could be useful for many experimenters. His system uses 3-D printed optical mounts for extruded aluminum angle, which can be purchased from most local hardware stores.[30] The aluminum angle design is shown in Figure 6.19. This system would be of particular use to anyone working in remote communities without shipping access to specialized extrusions like OpenBeam. Similarly, simple designs can also be coupled together to make much more complicated systems. Consider the

---

[29]Collection of 3D printable open-source optics equipment http://www.thingiverse.com/jpearce/collections/open-source-optics

[30]Optical Mounts http://www.thingiverse.com/thing:38062.

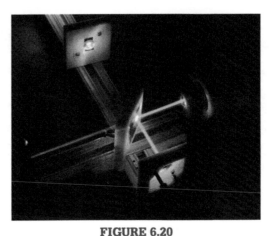

FIGURE 6.20
Michelson interferometer fabricated with 3-D printable kinematic mounts and extruded rails.

optics setup in Figure 6.20, where Thingiverse user ordaos uses several of the kinematic mounts shown in Figure 6.13 to make a Michelson interferometer.

Any researcher, scientist, or teacher, whether professional or amateur with access to a low-cost open-source 3-D printer, can utilize the designs to radically reduce the cost of optical support equipment as summarized in Table 6.1.

As can be seen in Table 6.1, cost reductions over 95% are common with some components representing only 1% of the current commercial investment. Even if the RepRap build discussed in Chapter 5 represents too steep of a learning curve or time commitment (~24h build time for one person, or 8h for two) than your laboratory or school can invest, dozens of small companies offer kits or prebuilt open-source 3-D printers, which are easily found on the Internet. With even assembled RepRap printers costing around US$1000 [8], printing a relatively simple optics setup or even a single filter wheel easily recoups the investment. The filament is also available on the web from dozens of suppliers (in various colors and polymer types) and our recent work has investigated the use of recycled polymer extruders (called "recyclebots"),[31] which decreases the cost of the material for the components by another order of magnitude [9,10].

The ability to custom manufacture optics equipment to specifications within a university, government, or industrial laboratory not only ensures that the components are exactly what you as the researcher need, but it also saves time. For example, it is much faster to print a predesigned component than to go to

---

[31]Recyclebot controls http://www.thingiverse.com/thing:54180 and mechanical system http://www.thingiverse.com/thing:12948

Table 6.1 Material and Energy Costs Associated with Open-Source Optics Component Fabrication Compared to Commercial Prices and Percent Savings

Components	Filament Consumption (g)	ABS Costs (USD)*	Electricity Cost (USD)§	Total Cost (USD)	Estimated Commercial Price (USD)¶	Percent Savings (com.-open)/ com.
Optical rail	–	–	–	10–12/m	320/m	97
Base on optical rail - Optical foot (2×) - Optical mag (3×) - Rod base (4×)	39.52	1.50	0.27	3.08	150–730	>97
Filter holder	8.98	0.34	0.06	0.40	58–80	>99
Lens holder	5.35	0.20	0.04	0.24	20–180	>98
Mirror holder	7.40	0.28	0.05	0.33	18–200	>98
Fiber switcher	10.41	0.40	0.07	0.47	22–138	>97
Screen holder	1.55	0.06	0.01	0.07	18	99
Thumb screw (6×)	7.98	0.30	0.06	1.32	12	89
Sample holder	6.00	0.23	0.04	0.27	18–109	>98
Lab jack	133.20	5.06	0.92	5.98	35–1000	83–99
Automated filter wheel changer	295.1	11.21	2.02	20.43	1000–4250	>98
Optical base (4×)+steel sheet vs optical table 1 m²	46.28	1.76	0.32	25.58	3619–5288	>99

*The price of 3mm ABS filament is $0.038/g. Source: 3D printer stuff. Available from: http://www.3dprinterstuff.com/shop/page/4?shop_param= [accessed 19.10.12].
§The national average cost of electricity is 11.53cents/kWh. Source: U.S. Energy Information Administration. Source: Available from http://www.eia.gov/beta/enerdat/#/topic/7?agg=0,1&geo=g&endsec=vg&freq=A&start=2008&end=2011&charted=1 [accessed 19.10.12]. The electricity cost derived from multimeter is 0.006925 kWh/g (using 3mm ABS) assuming a Prusa RepRap 3-D printer.
¶Commercial prices were derived from website data from various vendors including Edmund Optics, Thorlabs, McMaster-Carr, AutoMate Scientific, and Pasco.

a lab supply store if there is one in the area or even order online with next-day shipping. The value of timely access to experimental equipment can hardly be overstated for researchers as delays due to out-of-stock equipment and shipping are well-known problems for all researchers doing advanced experimental work. If a critical component is out of stock from suppliers, designing and printing it within a lab can literally save weeks. As the components of the open-source optics library are parametric, customizing the part and having a finished design ready for printing is extremely fast and easy. However, it should be pointed out that time savings are highly component specific. Three-dimensional printing in general is more time-consuming on a per-part basis than any mass manufacturing process, but there are savings associated with ordering, stocking and shipping for self fabrication. In addition, if a new component of significant complexity needs to be designed from the beginning, this can be more time-consuming than simply ordering a commercial component. However, if the component needs to be customized, self-fabrication again can be and is likely to be much faster and normally is far less expensive.

Although some of these optical components are less precise than commercial versions (as will be discussed in the next section), often experimental setups contain overly engineered components and the open-source tools described in this book provide more flexibility. For example, eliminating an optics table in exchange for using magnetic bases, a steel plate or repurposing an existing steel case desk, not only radically reduces the setup costs but also enables far more flexibility, time-saving and ease of reconfiguring an optical experimental setup.

It is important to note the benefit of this approach for international scientists. Experimental science is severely underfunded in most of the developing world as compared to Europe, the United States and Japan.[32] At the same time, the majority of humanity still lives in these regions and they possess talented scientists. Their training and theoretical background often rivals the preparation found in the West. Yet these scientists are severely handicapped by not having access to experimental equipment. This open-source approach would thus help enable more scientists to join the experimental side of the world scientific community, which I think is intuitively obvious to presume that it will be a benefit for everyone.

Finally, it should also be pointed out that there is a burgeoning collection of 3-D printing services, where users can upload a design and receive a 3-D printed part in the mail. These services can be used for more complex parts, components needing to be printed out of more advanced materials like metals, or for research programs interested in trying out the potential of 3-D printing for the science labs before embarking on 3-D printing themselves.

6.2.2. Benefits of the open-source approach for physics education

The open-source optics library provides an example of an easy way to design and conduct educational experiments while saving for schools tremendous amounts of money. Particularly, for teachers and lab instructors, planning multiple educational optical benches for classes using the open-source optics library would provide substantial savings. For example, to outfit an undergraduate teaching laboratory with 30 optics setups including 1 m optical tracks, optical lens, adjustable lens holder, ray optical kit, and viewing screen, the total cost would be <$500 using the open-source optics approach as compared to $15,000 for commercial versions, providing over $14,500 in savings. Thus,

---

[32]At the time of this writing, it appears the United States may be joining this category as well. Regardless, I doubt there are many working scientists that feel science is overfunded.

using the approach described in this book, the total cost can be reduced by an order of magnitude, allowing access to experimental setups in far more locations (e.g. countries with developing or transitional economies, economically depressed public school districts, and home schoolers) and levels of education (e.g. high schools, middle schools or even elementary schools). For some fortunate educational systems, the costs of the commercial physics educational equipment are trivial and are thus only a small factor in the decision-making process. Yet at many high schools in America, a teacher's yearly budget for supplies may only be $1000 for everything, and costs play a much more significant role. As it currently appears, the U.S. investment in education will continue to atrophy, the ability to reduce the costs of educational materials should not be understated even for the largest economy in the world. Even at fortunate institutions, money saved on open-source printable equipment can be invested in more sophisticated commercial equipment, faster computers, more teachers or teachers' aids, or other areas to improve student learning.

Besides economic savings, the ability to custom fabricate equipment with tolerable accuracy and precision for elementary-, middle- and high-school education in basic science, chemistry, and physics can save teachers time, similar to the benefits enjoyed by professional researchers. So for example, the rod holder could be easily adapted in OpenSCAD to the diameter of a birthday candle, again significantly reducing the cost of any light source, while saving the teacher the time in making a new design. With the open-source optics library, many middle-school, high-school and college-level experiments can be performed. The students could be responsible for creating the scales on paper secured directly under the rail, which could also be used to draw and label a ray diagram that includes the positions of the light source, lens and viewing screen. The students could learn to describe the image, real or virtual, upright or inverted, larger than object or smaller, gaining knowledge about the properties of the mirrors and lenses. As a useful exercise, students within the classes could be responsible for fabricating some of their own optics equipment with 3-D printers, or improving designs; this construction activity would cultivate students' practical abilities and exposes them to useful engineering skills (e.g. geometry, CAD and additive layer manufacturing) as well as the open-source philosophy.

Finally, it should be noted that there are substantial energy and environmental savings made possible by distributed manufacturing [10] and although optics equipment in general does not provide a relatively large environmental burden, it again is useful as a teaching tool for students when considering the design of other more substantively environmentally destructive types of equipment. Such discussions can help tie technical efficiency of energy conversion devices discussions in a physics class to what students are learning about in an environmental science course. Finally, as the RepRap and the recyclebot technology are highly portable and there have already been efforts to construct solar-powered 3-D printers,[33] it can be used to easy to support optics education in rural areas or developing countries and education institutions with significant funding constraints [11,12].

6.2.3.   Limitations and future work needed

There are several limitations of the open-source optics library described here. First, it is far from complete. There are hundreds of additional optical components that could be developed using this method that have not been designed yet. We grow the library as we need components from our work as do other labs working in the field, but there is much left to do. Please jump in and help! At the same time, there are many components that cannot be developed with the current technologies discussed here (e.g. lenses). There is considerable work in progress to make more functional printed materials and the idea of printing a nanoscale super lens or Fresnel lenses from transparent material is within the realm of near-future possibility. Second, there is sometimes a tradeoff between precision and cost, for both commercial optical equipment and the open-source home-made or lab-made variants discussed in this book. For example, the fundamental properties of the RepRap limit printed part resolution. The 0.5 mm Reprap extruder nozzle has a 2 mm minimum feature size, 0.1 mm positioning accuracy, and a layer of 0.2 mm thickness. There has already been considerable work done to move to smaller filament diameters and nozzle sizes, which have, for example, pushed the step height to 0.1 mm (100 μm), but extremely high tolerances are still not accessible to low-cost open-source 3-D printers. This will unquestionably change in the near future, but that does not help you now if this is needed for your project tomorrow.

In the same way, the Arduino platform is meant primarily for prototyping so microcontroller boards designed for a specific purpose can have better performance or have properties not available currently (e.g. able to handle higher voltage ranges). The Arduino platform is also costlier than just using the microcontroller chip itself, so there are always considerable cost savings from making a specialized board for a specific application such as the Melzi board discussed for the RepRap in Chapter 5. Third, in general, there are no warranties associated with open-source optics equipment—the users get what they make. Thus the quality of the components and the work that can be done are sometimes user dependent. For example, parts printed from acrylonitrile-butadiene-styrene (ABS), the same polymer that makes up Lego blocks, are relatively robust if printed with sufficient fill. However, the mechanical strength of the components is dependent on the quality of the print, which will vary among

---

[33]See, for example, our open-source solar-powered RepRap design http://www.thingiverse.com/thing:74671 and project page http://www.appropedia.org/Mobile_Solar_Powered_3D_Printer_V2.0.

printers/users. Further work is needed to determine if the layered material can endure consistent rough manipulations in educational applications (e.g. the purposefully destructive beatings equipment can take in public schools) so the lifetimes of printed parts can be compared to industrial-manufactured injection-molded components or those made in more robust materials such as steel. Finally, although the sharing of open-source optics designs significantly reduces the complexity of replicating equipment, there are still substantial knowledge sets necessary to take full advantage of the power offered by the open-source approach. These knowledge sets can act as barriers to entry for researchers and educators. For example, the software knowledge necessary to operate OpenSCAD, the printing software, and the Arduino coding is largely dependent on prior exposure to and basic programming ability and skills of the user. The more experience you already have in these realms can be leveraged to quickly complete highly advanced projects to develop scientific tools. Not all groups are necessarily as well endowed as some of the example labs we highlight throughout this book. However, there are a vast array of free online tutorials, videos, examples and instructional materials available for the novice users for all three types of software. In addition, there is work by our group and others (e.g. the entire education system of the Netherlands) to assist young students gain direct exposure to open-source 3-D printing and Arduino programming. As these students climb up the ranks of academia, they will represent an enormous wealth of super user/developers to accelerate the evolution of scientific equipment and science itself in both academic and industry labs.

Future work on the technological development of this open-source optics model is necessary to meet the full potential of the concept. Although 3-D rapid prototyping is currently used primarily in research and development and thus contributes only to a tiny fraction of global manufacturing, the process has an enormous potential to fabricate more complex components with improved precision and materials selection. RepRap-like printers need to be developed that can print other materials with sufficient resolution to produce lenses, filters, mirrors, and other both optics and nonoptics equipment in the physics lab. With advanced deposition techniques, chemically active components and optical coatings could be printed. Thus mirrors and filters with different wavelength ranges could be custom digitally fabricated by simply depositing desired species. In addition, 3-D printers in the future are expected to have higher resolutions, which enable other applications. Taken to the atomic limits, 3-D printing can be applied to nanofabrication, nanoimprinting, and nanoscale deposition techniques that open up further applications. With strong cases being made for more open-source approaches in nanotechnology [13,14], the codevelopment of these concepts could provide enormously powerful tools. For example, a microscale 3-D printer could print the integrated circuit used in open-source optics or a nanoscale 3-D printer can print the gratings used for light diffraction. In addition, 3-D printers and nanotechnology provide the opportunity to fabricate digital designs of chip-like optical systems further depressing costs of experimental equipment and opening the possibility of making them as ubiquitous as cell-phone cameras.

The methodology described here can also be used to make more advanced optical devices such as spectrometers (which we will look at in the last section of this chapter), monochromators and ellipsometers and of course other areas of physics. Not only optical apparatuses can be built in this way; mass spectrometers, chromatography and even X-ray diffraction systems and other equipment are also theoretically printable to a large extent using next-generation open-source 3-D printers. On the other hand, as the 3-D printer itself is reduced in size, it is also possible to have built-in 3-D printers inside the large machines, serving as an in-situ assistant for components replacement, circuit reparation and in-situ design and in-situ fabrication. Symes et al. have already reported the application of 3-D printer as reactionware for chemical synthesis and analysis [15]. This enabled reactions to be initiated by printing the reagents directly into a 3-D reactionware matrix and to be monitored in situ. The construction of a relatively cheap, automated and reconfigurable chemical platform makes the techniques from chemical engineering accessible to traditional synthetic laboratories [15].

A large number of open-source software programs and open-source databases have been built in recent years that benefit scientists [16–26]. As OSH becomes more mainstream and open-source data sharing allows everyone everywhere at any time to design, build, share and comment on OSH, the utility of the approach will create a virtuous cycle. As people design, build, share and comment, they contribute value to the open-source communities, which everyone again can benefit from and thus encourage more participation. As this section has demonstrated, this has already started in the field of optics, can spread throughout the rest of physics and the other sciences and is applicable to most fields.

6.2.4. Section summary

This section introduced a library of open-source 3-D printable optical components to provide an extremely flexible, customizable, low-cost, start of a public-domain library for developing both physics research and teaching optics hardware. Using this open-source optics method can reduce costs of many optical components by 97% or more. It is clear that this method of scientific hardware development enables a much broader audience to participate in optical experimentation as both teaching and research platforms than previous proprietary methods.

6.3. ENGINEERING: OPEN-SOURCE LASER WELDER, RADIATION DETECTION, AND OSCILLOSCOPES

6.3.1. Open-source laser welder

One of the most useful facets of open-source equipment is the ability to build on one anothers' work. This section describes the development of an open-source automated polymer welding system for testing heat exchanger designs, which was built on the open-design computer numerical control (CNC) laser community.

Peter Jansen[34] was a graduate student in the Cognitive Science Laboratory at McMaster University in Canada. Among his many projects and accomplishments, he became interested in 3-D printing and the RepRap project discussed in detail in Chapter 5. He decided to take the RepRap project to the next level and began designing and constructing prototypes for an inexpensive 3-D printer based on selective laser sintering (SLS), which has the advantage of allowing the creation of complex arbitrary geometries, with potentially much higher resolution than fused filament deposition used in conventional RepRaps. In addition, in the long run, SLS has the potential to lower overall costs and eliminate many of the known issues surrounding the construction of durable and robust-fused filament extruders. The goal of his project is to construct an open-source laser cutter with a large cut area (about $1\,m^2$), for about 5–10% of the cost of a commercial system.[35] The design[36] shown in Figure 6.21 draws heavily from previous open-laser cutter projects, such as the Buildlog 2X Laser Cutter[37] in using inexpensive aluminum extrusion and optics for most of the structural frame.

In the Jansen design, however, the custom parts are 3-D printed from ABS (again radically reducing the cost of machining custom parts). The printed parts represent about 10 h of total printing time on RepRap and include parts such as NEMA17 motor holders that mount onto T-slot, idler brackets, pillow block bushing mounts for motors, idlers, and shafts.

His project follows the free and open-source model outlined in Chapter 2 and he posted designs for an open-laser cutter on Thingiverse. Even better, the designs were made with the hope that they would be of general utility to anyone printing out a large CNC system—not just a laser cutter. This aspect is the most important for my group's research, as we suddenly developed the urgent need to be able to weld thin sheets of plastic together. Here is why.

---

[34]Peter Jansen http://www.tricorderproject.org/aboutpeter.html.

[35]Commercial SLS systems can easily cost more than $500,000, so that still provides a lot of capital for a good laser source and optics.

[36]3D-Printable Laser Cutter http://www.thingiverse.com/thing:11653.

[37]Buildlog 2× Laser Cutter http://www.buildlog.net/blog/2011/02/buildlog-net-2-x-laser/.

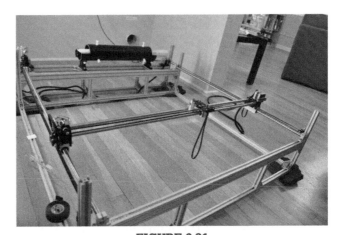

FIGURE 6.21
Open-source laser cutter.

In collaboration with David Denkenberger, Michael Brandemuehl, and John Zhai at the University of Colorado, I had demonstrated the ability to make an absurdly low-cost polymer-based heat exchanger from plastic using an expanded microchannel design [27]. Polymers, in general, have poor thermal conductivity and are not the obvious first choice for a heat exchanger material. Most current heat exchangers use metals and although microchannel heat exchangers are currently used, and they have low material costs, the manufacturing techniques are prohibitively expensive for most applications. Fortunately, although the thermal conductivity of polymers is generally orders of magnitude lower than metals, as long as the polymer walls are made thin, the thermal resistance is negligible. So we made the walls thin enough to enable high efficiencies to be possible using garbage bags. We measured an effectiveness of a prototype with 28-μm-thick black low-density polyethylene walls and counterflow, water-to-water heat transfer in 2 mm channels. It was a respectable 72% and multiple low-cost stages would provide the potential for very high effectiveness. This was simply awesome! We were able to get a garbage bag heat exchanger to outperform metal heat exchangers on the market that cost hundreds of dollars. The potential for the technology is enormous as heat exchangers are everywhere—refrigeration cycles, heat recovery, industrial processes, vehicles, and conventional power plants. In addition, the potential to have low-cost heat exchangers opens up applications no one has ever considered because of costs, such as ultra-low-cost appropriate technology for development [20]. In fact, David Denkenberger and I originally became interested in heat exchanger design when we developed a simulation of a solar water pasteurization system [28]. Our simulation showed that the lowest cost safe drinking water could be made with a solar water pasteurizer if we could get the

FIGURE 6.22
Open-source polymer welding system derived from open-source laser cutter in Figure 6.21.

total system cost under $25. We needed a good heat exchanger to make it work and those were expensive (e.g. starting at $250). I vividly remember walking through a rough part of inner city Philadelphia awkwardly carrying hundreds of dollars of conventional heavy-metal heat exchangers and nervously hoping I was not mugged. I made it safely, and in retrospect, my worries were probably unfounded as there are not a lot of street thugs that would be able to put an accurate value on heat exchangers. It was clear that any form of conventional heat exchanger design was simply never going to work in the real world for this application, which demanded a radical approach to cost. The expanded polymer heat exchanger was just what we needed, we had proved it could work, but there was one catch. The prototyping costs to make the heat exchangers despite the material costs being almost free—ran hundreds to thousands of dollars for the necessary precision polymer laser welding. Enter Jansen's open design.

To develop a polymer laser welding system from scratch would have been extremely time-consuming and difficult, but to build on free and open-source designs is much faster and involves less effort. A group of my students including Rodrigo Periera Faria (from Brazil), Sai Ravi Chandra Parasaram (from India) and from the U.S. Nick Anzalone and Thad Waterman used Jansen's design as a starting point to develop an open-source polymer welding system as shown in Figure 6.22.

The primary variations we made to Jansen's system with dimensions of 1.5 ft by 1.5 ft included the following:

1. We mounted the whole system in a double drawer file cabinet (now that we all more or less only read academic papers on a screen, there are

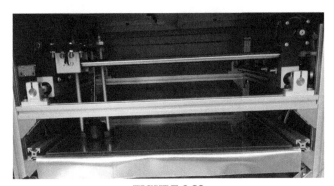

FIGURE 6.23
Open-source polymer welding system detail.

millions of filing cabinets available throughout the world's universities) that we hollowed out, put on a safety switch for the laser, and new drawers with magnetic closers. A hole was drilled through the top for the wires and the fiber laser to pass through.

2. We extended the legs past a square bottom to allow for the cabinet rails to remain intact as seen in Figure 6.22.
3. We added a second substrate layer that slides in consisting of a metal (Al) substrate and a lower iron glass cover plate (see details in Figure 6.23).
4. We added new 3-D printed parts which are available on Thingiverse[38] to couple a fiber laser to the rig. In our case, it is hanging down just over the glass with a lens positioned so that the focal point is just under the glass.

In addition to the printed parts, the system needs mechanical and electrical parts shown in Table 6.2. Finally, we used a fiber laser with full details available on Appropedia. It should be noted here that, like most open-source designs, this is a living, evolving research tool and any changes will be posted on Appropedia as we improve our design.[39] To control the system, we used the electronic circuit schematic shown in Figure 6.24. The electronics build and communication setup has 10 easy steps:

1. *The jumper near to the GND pin must be removed*; otherwise the Arduino will be damaged! (See details in Figure 6.25).
2. Connect the 12 V cable to M+ pin and ground to GND pin on the motor shield as seen in Figure 6.26.
3. Connect the endstops as shown in Figure 6.27.

---

[38]Open-source laser polymer welding system http://www.thingiverse.com/thing:28078.
[39]Open-source laser polymer welding system wiki page http://www.appropedia.org/Open-source_laser_system_for_polymeric_welding.

Table 6.2 Parts for the Open-Source Polymer Laser Welding System

Misumi Part Description	Part Number
Precision linear shaft	PSFJ12-480
Linear double bushings with pillow blocks	LHSSW12
Linear single bushings with pillow blocks	LHSS12
Aluminum extrusion four-side slots	HFS5-2020-2000
Square nuts for aluminum extrusions	HNKK5-5
Reversal brackets with tab	HBLFSN5
Cap screws for aluminum extrusions	HCBST5-12
T-shaped shaft supports	SHA1220
Stock Drive Part Description	**Part Number**
18 teeth polycarbonate timing pulley	A 6T16M018DF6005
Fiberglass-reinforced neoprene toothed pulley belt	A 6Z16MB89060
Electronic Parts	**Source/Part Number**
Arduino MEGA 2650	Sparkfun 11061
Adafruit Motorshield	Adafruit 81
2 × Adafruit stepper motor	Adafruit 324
2 × Opto endstop	Makerbot
Power supply 5 V and 12 V	(Used computer)
MAX3323 chip (or similar)	Digikey
4 × 1 µF capacitors	Favorite electronic shop
Female serial-breadboard cable	Favorite electronic shop
Hookup wire	Favorite electronic shop

4. Fix the endstops such that the laser carriage reaches the lower left corner of the mechanism as shown in Figure 6.28.

5. For the following steps, use the chip and wiring pins outline in Figures 6.29–6.31. If using Adafruit stepper motors, connect the wires following the order: from M1 to M2, brown, green, skip Gnd pin, yellow and red. From M4 to M3, brown, green, skip Gnd pin, yellow and red.

6. Connect MAX3323E pin 1 to a 1 µF capacitor, a 1 µF capacitor between 1 and 3, another between 4 and 5, and the last 1 µF capacitor to pin 6, as shown in the figure. Pins 7 and 8 go to the DB9 cable pins 3 and 4, pins 11, 12, 13, 14 and 16 to 5 V and pin 15 to ground. Pin 9 to Arduino pin 19 and pin 10 to Arduino pin 18. Ground the capacitors as shown, if using polarized capacitors, make sure to connect the negative sides to ground for the capacitors in MAX3323E pins 1 and 6, pin 3 for the capacitor between 1 and 3 and pin 5 for the capacitor between 4 and 5.

7. Connect DB9's pins 1 to ground, pins 2 and 5 to pin 9, pin 3 to MAX3323 pin 7, pin 4 to MAX3323 pin 8, and pin 7 to pin 8.

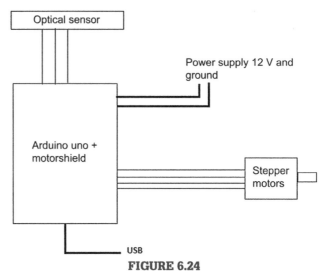

FIGURE 6.24

Electronic circuit schematic for the open-source polymer welding system.

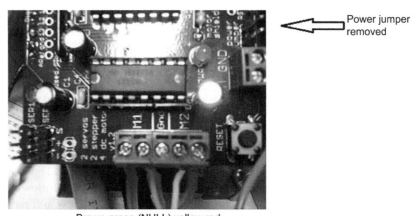

Brown green (NULL) yellow red

FIGURE 6.25

The jumper near the GND pin must be removed, otherwise the Arduino will be damaged!

8. Upload the firmware to Arduino using Arduino software.[40]
9. Connect the DB9 cable to LaserSource 4320 RS232 input.
10. For firmware, open-source software, and a template for this tool go here:
    https://sourceforge.net/projects/lasersystemforp/.

---

[40]Details in using Arduino's are in Chapter 4. Everything you need is available http://www.arduino.cc/

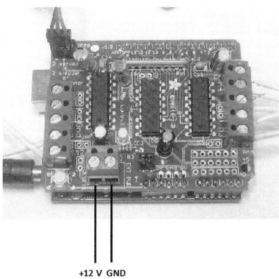

FIGURE 6.26

Connect the 12V cable to M+ pin and ground to GND pin on the motor shield.

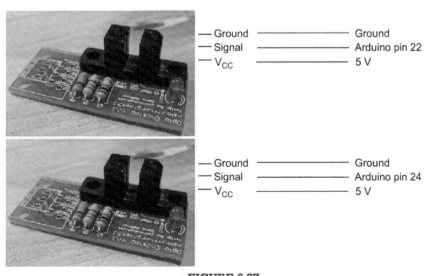

FIGURE 6.27

Connect the optical endstops.

FIGURE 6.28

Placement of endstops to set system boundaries.

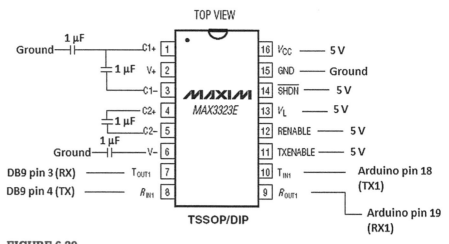

FIGURE 6.29

MAXIM 3323E pin assignment for open-source polymer welding system.

Next, set up the laser, which is composed of three main devices[41]: (1) Laser-Mount 264, (2) TECSource 5300 and (3) LaserSource 4320. The LaserSource 264 integrates a Peltier cooler for precise temperature control and the laser itself. TECSource 5300 is a temperature controller that needs to be attached

---

[41]These devices are not open source. Many times, even when developing open-source hardware for your laboratory, until the OSH field is more mature, specialized closed components will need to be acquired and integrated into a solution. As the OSH community develops, this will become less necessary and the costs of these hybrid types of scientific tools will be further reduced.

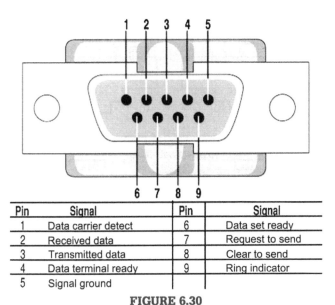

Pin	Signal	Pin	Signal
1	Data carrier detect	6	Data set ready
2	Received data	7	Request to send
3	Transmitted data	8	Clear to send
4	Data terminal ready	9	Ring indicator
5	Signal ground		

FIGURE 6.30

DB9 female cable front view.

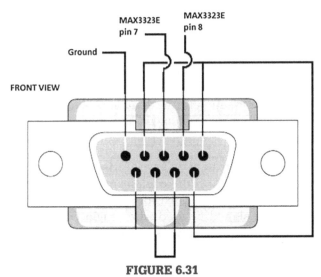

FIGURE 6.31

DB9 female cable pin assignment.

with the LaserMount. LaserSource 4320 is a Laser Diode Driver; it controls the laser behavior such as voltage, current, pulse width modulator (PWM) duty cycles and on/off control.

Installation of the LaserSource and the TECSource are fairly straightforward. After unpacking the units, make sure all packing materials have been removed and nothing obscures the ventilation ports on the side and front of the units. Change the voltage selection to the appropriate value and make sure both devices are properly grounded. The devices have vent holes on the side and front; do not block these vent holes, or overheating may occur, causing damage to the unit. Connect the cables from the TECSource- and the LaserSource- labeled LASER and TEC to the LaserMount properly. To power up the unit, connect the AC power cord to the unit, turn the power switch, located on the front panel, to the on (I) position. The unit will display the model, serial number, and firmware version, go through a quick power-up self-test, and return to the last-known operating state. In order to achieve the highest level of accuracy, the TECSource should be powered on for at least 1 h prior to taking measurements. Once devices are powered up, it is necessary to enable the External Fan Control on the TECSource menu options. Make sure the temperature controller current limit is set to a maximum value of 7.4 A.

As with all lasers, you need to take proper safety precautions with this setup. Below are some general guidelines, but you should contact the manufacturer and your institution's safety personnel before you build or operate this system. At your institution, there are probably rules governing laser safety in detail. At our institution, the LSO needs to be notified of the purchase of any laser, regardless of the class. Such notification should include the classification, media, output power or pulse energy, wavelength, repetition rate (if applicable), special attachments (frequency doublers, etc.), beam size at the laser aperture, beam divergence and users. We overcame the requirement to keep the laser in a shielded room by putting in a box with a safety switch, which does not allow the laser to turn on if the doors are open. To make this effective, follow this primary rule: *when operating the laser, make sure that the doors to the rig are closed and that you are wearing laser safety glasses.*

No attempt should be made to place any shiny or glossy object into the laser beam other than that for which the equipment is specifically designed. Eye protection devices which are designed for protection against radiation from a specific laser system should be used when engineering controls are inadequate to eliminate the possibility of potentially hazardous eye exposure (i.e. whenever levels of accessible emission exceed the appropriate MPE levels). This generally applies only to Class IIIB and Class IV lasers. All laser-protective eyewear should be clearly labeled with optical density values and wavelengths for which protection is afforded.

Skin protection can best be achieved through engineering controls. If the potential exists for damaging skin exposure, particularly for ultraviolet lasers (200–400 nm), then skin covers and or "sun screen" creams are recommended. Most gloves will provide some protection against laser radiation. Tightly woven fabrics and opaque gloves provide the best protection. A laboratory jacket or

FIGURE 6.32
The aluminum + glass slider.

coat can provide protection for the arms. For Class IV lasers, consideration should be given to flame-resistant materials.

The software for this system is all free and open source and includes the following:

1. Hydra—MMM modified for the Laser Welder.[42] Java JRE required.
2. Processing[43]—if using source code.
3. Inkscape[44]—for making new laser path designs. You will also need the GCodeTools for Inkscape.[45] To install the Gcode tools in Linux unpack and copy all the files to the following directory, /usr/share/inkscape/extensions/ and restart Inkscape.
4. You can also get the template file from Appropedia.[46]

To load a sample for a welding experiment, follow these steps:

1. Open the safety doors.
2. *Carefully* remove the aluminum + glass slider as shown in Figure 6.32.
3. Remove the glass.
4. Place the plastic layers (two the first time and adding one per weld routine if more than two).

---

[42]Hydra http://sourceforge.net/projects/lasersystemforp/.
[43]Processing http://processing.org/.
[44]Inkscape http://inkscape.org/download/?lang=en.
[45]GCodeTools for Inkscape http://www.cnc-club.ru/forum/viewtopic.php?t=62#top1.
[46]Template example http://www.appropedia.org/File:MOST_laserweld_TEMPLATE.svg.

5. Stretch the plastic layers flat/cold weld with rolling pin and make sure that they are touching each other with as few wrinkles as possible. Alternatively, you can pin them down over the edges.
6. Put the glass on it.
7. Slide the arrangement back to the box carefully to not break the laser mount (see details in Figures 6.33 and 6.34).
8. When the substrate is mounted properly as seen in Figure 6.35, *close the box and put on the safety glasses.*

FIGURE 6.33
Rail slider detail.

FIGURE 6.34
Be careful! The laser rig rides just above the glass.

Next, to power up the open-source laser welder and the associated communications, do the following:

1. Disconnect USB cable as shown in Figure 6.36.
2. Turn on the power supply shown in Figure 6.37 (*Do not connect the USB cable before this!*).
3. Connect the USB cable.
4. Turn on TECSource 5305 (shown in Figure 6.38), wait until it boots, and then press Output.

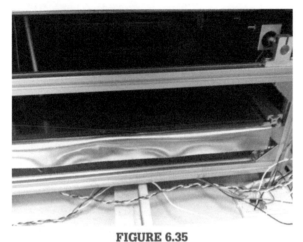

FIGURE 6.35
The substrate is mounted properly and ready to process.

FIGURE 6.36
USB cable.

FIGURE 6.37
Power supply.

FIGURE 6.38
TEC source.

5. Wait until it stabilizes at the temperature set point (currently 25 °C).
6. Turn on 4320 LaserSource as shown in Figure 6.39.
7. Open the Hydra GUI and click on "connect" as shown in the screenshot in Figure 6.40.
8. If you receive this error message (e.g. Figure 6.41), restart the program and reset the Arduino (Figure 6.42), and then click on "connect" again.
9. Connected (Figure 6.43) and ready for use!

Next, you will want to make a design to weld. This is done by drawing the design in Inkscape and exporting GCode (similar to the slicing that takes place

FIGURE 6.39
Laser source.

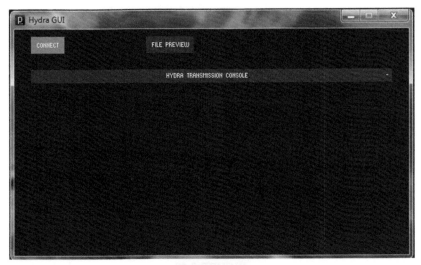

FIGURE 6.40
Hydra GUI.

for the RepRap in order to tell the tool head where to go). The process is as follows:

1. After installing Inkscape and GCodeTools following the instructions above, run Inkscape.
2. Open the file TEMPLATE.svg as shown in Figure 6.44. The template has the working area limits and other things necessary to generate the GCode correctly for this specific design of open-source welding rig.

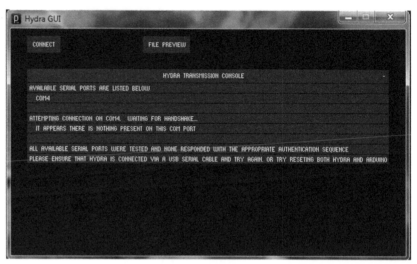

FIGURE 6.41
Hydra GUI connection problem.

FIGURE 6.42
Arduino green light.

3. Draw a line, as shown in Figure 6.45, by selecting the line and press Ctrl + Shift + C, as shown in Figure 6.46. In the same way, you can build up much more complicated designs.
4. Go to Extensions >> Gcodetools >> Path to Gcode (as shown in Figure 6.47).
5. On the tab Preferences, type the File name (with .gcode in the end) and the directory you want to save it in.
6. Go to Path to GCode tab and Apply. The GCode should now be generated to the directory folder.

FIGURE 6.43
Connected and ready for use.

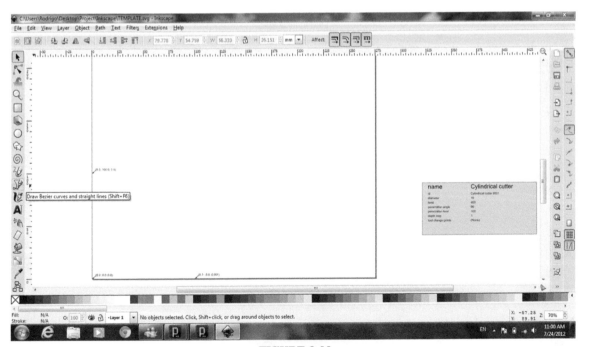

FIGURE 6.44
Inkscape template.

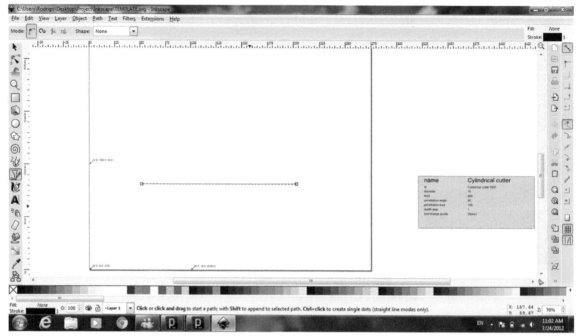

FIGURE 6.45

Line in Inkscape.

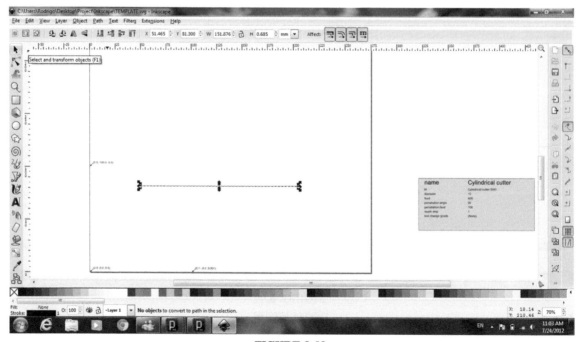

FIGURE 6.46

Selected line in Inkscape.

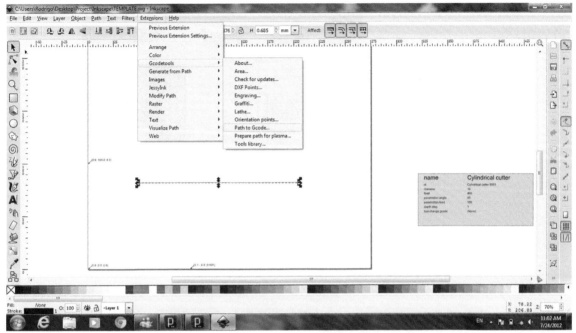

FIGURE 6.47
Path to Gcode.

Now, you are finally ready to weld!

1. *Close the box and put on the safety glasses.*
2. Turn the LaserSource enable key to *On* (Figure 6.48).
3. In Hydra, click on Send File, select the GCode file and click open again (Figure 6.49).
4. Wait until the message END OF PROGRAM appears.
5. Turn the enable key back to *Off*, and open the box.

Now be amazed at your perfect polymer weld as shown in Figure 6.50.[47]

For full operations and any changes/improvements and new versions, see the MOST protocol page.[48] Designing the open-source laser welder in this way and sharing everything associated with it not only makes it easy for future students in our group, like Brennan Tymrak, to start large-scale proto-typing and testing of expanded microchannel polymer heat exchangers, but

---

[47]You are, of course, unlikely to get a perfect weld the first time and you will need to do some trial and error investigation into your hardware variables (e.g. laser optics, motor speed, time and intensity of laser power, etc.).

[48]Protocol page http://www.appropedia.org/Laser_welding_protocol:_MOST

FIGURE 6.48
Laser source enable key to turn on.

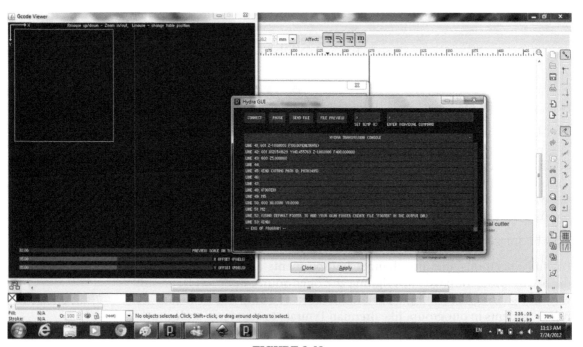

FIGURE 6.49
Running the code.

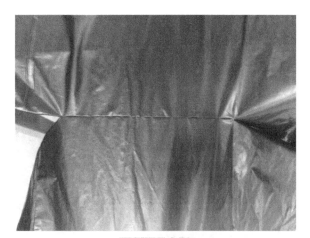

FIGURE 6.50
Resultant polymer laser weld.

it also makes it easy for others to build on our technologies. The benefits of this sharing approach for an academic are well known, as discussed in Chapter 2.

6.3.2. Radiation detection with open-source Geiger counters

There are many scientists and engineers that work with radioactive materials and may find it useful to have a low-cost, reliable, open-source radiation detector. In addition, many "citizen scientists" may be interested in obtaining reliable first-hand information about dangers near a nuclear power plant accident or other forms of radioactive release.

There are several OSH Geiger counters available. For example, the design of the Libelium Team's board,[49] which is a shield for the Arduino (discussed in Chapter 4), is open hardware and the source code is released under GPL (discussed in Chapter 2). In addition, there is an open Geiger counter circuit[50] that interfaces with an Arduino and a project kit to build it is available.[51] The Geiger Kit provides the electronics as shown in Figure 6.51 (and assembled in Figure 6.52) needed to run and detect events from a GM tube.

---

[49]Geiger counter—Radiation Sensor Board for Arduino http://www.cooking-hacks.com/index.php/documentation/tutorials/geiger-counter-arduino-radiation-sensor-board.
[50]Geiger counter circuit https://sites.google.com/site/diygeigercounter/circuit-description.
[51]DIY Geiger counter kit https://sites.google.com/site/diygeigercounter/home.

FIGURE 6.51
Geiger counter kit Gk-B.

FIGURE 6.52
Geiger counter kit GK-B5 assembled.

The events are counted and displayed as counts per minute and microSievert per hour (μSv/h) as shown in Figure 6.53 by an ATmega328P microprocessor running preloaded software. The data can be output to a computer via a serial connection.

With the use of a 3-D printer as discussed in the last chapter, you can even print a case with a detachable probe to house your Geiger counter with a design by Stephen Bailey, shown as designed in Figure 6.54 and fully assembled in Figure 6.55.[52]

---

[52]Case designed for the DIYGeigerCounter v.4.1 http://www.thingiverse.com/thing:36678.

FIGURE 6.53
Geiger counter kit display.

FIGURE 6.54
3-D printable case for Geiger counter with a detachable probe.

For such projects, a major impetus was to help people in Japan to measure the levels of radiation in their everyday lives after the unfortunate earthquake and tsunami struck in March 2011 and caused nuclear radiation leakages in Fukushima. Due largely to conflicts of interest, the official announcements of nuclear accidents are viewed as unreliable by much of the public.[53] For

---

[53]Consider Pennsylvania's Three Mile Island nuclear incident several decades ago. There is now substantial evidence that the releases were underreported to the public by officials by at least an order of magnitude. The official Nuclear Regulatory Commission (NRC) value is 10 MCi [57]. Thompson et al. quote more than double that at 22 MCi [58], whereas Gundersen points out that the sum of the NRC releases yields 36 MCi and estimates anywhere between 100 and 1000 times the NRC value [59]. The bottom line perhaps comes from a raft of epidemiological studies, which point to a significant epidemic of cancer that is clearly related to the Three Mile Island release and that would not have occurred if the official NRC values were accurate [60–66].

FIGURE 6.55
Fully assembled Geiger counter.

example, during the Fukushima disaster, the Japanese public found official reports extremely dubious and government officials appeared to be actively preventing citizens from obtaining data [29]. Even in the United States, our government appeared reluctant to post online whatever radiation levels they were monitoring as radiation from Fukushima hit the West Coast, and there were several reports that their monitors crashed [30]. A response from citizen scientists in Japan was to crowd-source radiation Geiger counter readings from across their country using a collection of both OSH and open-source software [31] as can be seen in the Japan Geigermap: at-a-glance both then (Figure 6.56) and now (Figure 6.57).

This kind of activity may be the start of a movement to bring more people into science and act as a check on bad or dishonest sources of scientific information for the public [32]. Conceptually, this type of crowd sourcing of scientific data is considered as part of the "Participatory Sensing" framework, whereby ad-hoc sensor networks are formed, taking advantage of sensors (possibly embedded in mobile devices[54]) to enable public and professional users to gather, analyze and share geographically localized information [33]. Scientists have already concluded from the many crowd-sourced experiments completed that individuals can act as sensors to give useful results in a timely manner and can

---

[54]There is already an Android App that can turn your cell phone into a radiation detector using the CMOS camera covered by a piece of tape http://hackaday.com/2012/01/15/turn-your-camera-phone-into-a-geiger-counter/ and Softbank launched a smartphone that tracks radiation specifically for Japan http://news.cnet.com/8301-1035_3-57444283-94/softbanks-geiger-counter-smartphone-start-of-a-global-trend/.

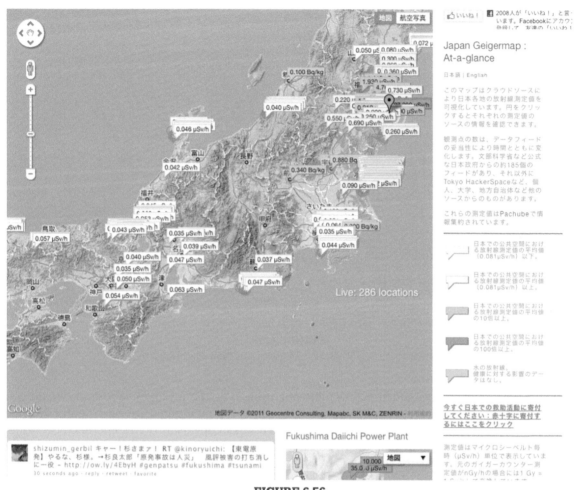

FIGURE 6.56

Screenshot of Japan Geigermap: at-a-glance during the Fukushima disaster.

complement other sources of data to enhance our situational awareness and improve our understanding and responses to such events [34].

6.3.3. Xoscillo—an open-source oscilloscope

As you are building your OSH for other projects, you may need to debug an electronic circuit. Although a simple multimeter is an indispensable tool for doing this, as your projects become more complex, you may need a more sophisticated tool to understand what is going on in the circuit in the form of wave forms. This calls for an oscilloscope. The least expensive commercially available oscilloscopes cost around a few hundred dollars. What if you could

FIGURE 6.57

Screenshot of Japan Geigermap: at-a-glance as of the writing of this book.

have functional oscilloscope using only the inexpensive Arduino board you already have? You are in luck. *Xoscillo*[55] is an inexpensive, open-source software oscilloscope that acquires data using an open-source Arduino microcontroller board (discussed in Chapter 4) or a Parallax USB oscilloscope. It has been released under CC-BY-NC-SA. The Xoscillo does not have all of the functionality of the high-end oscilloscopes, but it may meet many of your needs. There is no need for extra hardware to get the basic functionality, but the maximum frequency is 7 kHz. You can obtain up to four channels, but this lowers your sample rate (7/4 = 1.75 kHz). It has 8-bit vertical resolution, variable trigger voltage on channel 0, and can sample data indefinitely.

The hardware of the simplest embodiment of the Xoscillo is just an Arduino board with its analog inputs connected to the output of the circuit where the waveform is that needs to be observed. This setup is connected to a PC, and the inputs read by the Arduino are analyzed and shown on the screen of the PC.

To get started, flash your Arduino with the firmware that comes in the zip file at the link above. Use the analog inputs from #0 to #3 on your board. Launch

---

[55]Xoscillo https://code.google.com/p/xoscillo/.

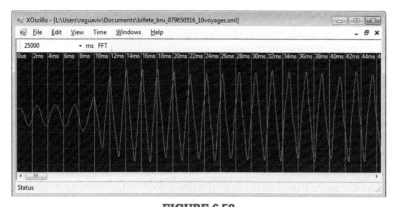

FIGURE 6.58
Screenshot showing the Xoscillo displaying a simple waveform.

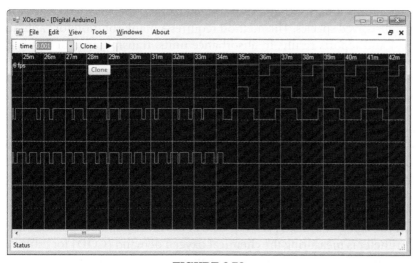

FIGURE 6.59
Screenshot showing the Xoscillo acting as a logic analyzer.

the application and in the menu, go to "File" and then click on "Arduino". A basic screenshot showing the Xoscillo displaying a simple waveform and a logic analyzer is shown in Figures 6.58 and 6.59, respectively.

6.4. ENVIRONMENTAL SCIENCE: OPEN-SOURCE COLORIMETERS AND PH METERS

OSH can also be used for environmental science. In this section, we will look at designs for an Arduino-based colorimeter and pH meter. After detailing the design and fabrication of the devices, the quality is discussed to provide data for conclusions about future development of other open-source analytical tools discussed in Chapter 7.

6.4.1.   Open-source colorimeter

Colorimetry is a scientific technique that is used to determine the concentration of colored compounds in solutions by the application of the Beer–Lambert law, which states that the concentration of a solute is proportional to the absorbance. To do colorimetry, you need a colorimeter, which starts in the hundreds of dollars for a lab-grade tool. In general, a *colorimeter* is used to measure the absorbance of only a particular color (e.g. wavelength) of light for a specific solution. A colorimeter is a relatively simple scientific device consisting of a light source, sample holder (normally a cuvette or test tube), light intensity sensor and means of controlling the light source and integrating transmitted light intensity. Incident light is generally filtered allowing only a narrow band of wavelengths near the absorbance peak for a given dissolved species. The method requires a blank solution for calibration (zero) and reports results in absorbance units, transmittance or if it is calibrated, it can apply the Beer–Lambert law to report results as a concentration [35]. Colorimetric methods are used widely in research and in many industries, including investigating the food we eat—such as during storage of bread [36], chocolate [37] and milk [38]. For example, in environmental science, colorimeters are used to monitor the levels of nitrates, phosphates, metals and other compounds present in effluent entering the natural environment from the manufacturing of paper and other commercial goods [39,40]. They have also traditionally been used to estimate the population density of protozoa in a culture, and more recently by the medical community, to measure the UV radiation exposure of children though skin color changes and to study the aging of bruises [41–43]. One of the more popular applications of a colorimeter is measuring the chemical oxygen demand (COD) for estimating the organic content of wastewaters. This is particularly important for some areas of environmental science. COD is also a measure included in some water quality indices [44]. There are sophisticated and expensive methods to determine COD with high accuracy [45,46], but often at high cost and increased production of waste from the analyses.

This section provides a design and technical validation of an open-source colorimeter [47] using the closed reflux COD method (EPA method 5220D). This approach is evaluated for its potential to reduce the cost of equipment to perform colorimetric COD.

The open-source colorimeter case design was wholly completed in OpenSCAD following the techniques outlined in Section 6.1.1. The design of the case body

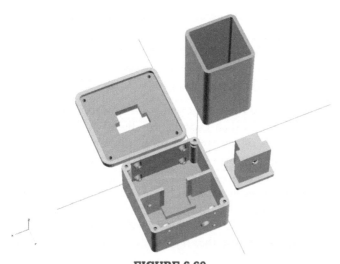

FIGURE 6.60
Schematic of case design in OpenSCAD of the open-source colorimeter.

FIGURE 6.61
Assembled case with electronics of the open-source colorimeter.

is shown schematically in Figure 6.60 and is available on Thingiverse[56] and is discussed in detail including bill of materials (BOM) and firmware on Appropedia.[57] The assembled case with electronics is shown in Figure 6.61.

---

[56]Open-source colorimeter STL and SCAD files http://www.thingiverse.com/thing:45443.
[57]Open-source colorimeter wiki page http://www.appropedia.org/Open-source_colorimeter.

The case was printed with black polylactic acid (PLA) media with three outer layers so as to minimize stray light inside the detection area. An open-source RepRap 3-D printer following the design outlined in Chapter 5 can be used to produce the print after slicing the OpenSCAD-produced STL model with the open-source slicing software such as Slic3r or Cura.

The electronics are based upon the open-source Arduino prototyping platform discussed in Chapter 4, which exposes the digital and analog I/O as well as processing capabilities of an Atmel Atmega microprocessor in a single convenient package. The platform encapsulates the hardware required to interface the Arduino with a host computer and includes a custom bootloader that executes compiled C++ code that can be developed in the Arduino integrated development environment (IDE). As we discussed, the Arduino platform is designed to use customized electronic boards (shields) that can be conveniently pressed into place and that typically come with software libraries so as to facilitate integration of board features into the custom code developed by the end user. A display shield incorporating a 16×2 character alphanumeric liquid crystal display (LCD) and D-pad button interface is used to navigate and select device functions and display the results of analysis. The shield is also OSH, supplied by Adafruit Industries,[58] a company dedicated to developing and providing low-cost, innovative and useful open-source electronic solutions. A single 5 mm through-hole LED having an emission peak near 606 nm was used as a light source (this of course can be exchanged for an LED of another wavelength for your specific application). A Taos TSL230R light intensity-to-frequency sensor was employed to measure incident light intensity. A total of only three discrete electronic components are required for the circuit, as shown schematically in Figure 6.62.

The schematic was developed at Fritzing.org,[59] which is extremely useful for sharing your electronic design schematics. It is clear from Figure 6.62 that the design does not come close to need the full functionality of even the Arduino Uno and thus could be made even more inexpensively. This is left for future researchers working on specialized projects because again using the Arduino greatly accelerates the ability of the user to get a functioning device in a short amount of time for a wide range of applications. The open-source colorimeter's firmware was developed with the Arduino IDE by G. Anzalone and provides an easy-to-navigate hierarchical menu system for selection of the device's functions.

---

[58]Adadfruit https://www.adafruit.com/.
[59]Fritzing is an open-source hardware initiative to support designers, artists, researchers and hobbyists to work creatively with interactive electronics. Fritzing has created (and is improving) a software tool, a community website and services in the spirit of Processing and Arduino, fostering an ecosystem that allows users to document their prototypes, share them with others, teach electronics in a classroom, and layout and manufacture professional printed circuit boards (PCBs). http://fritzing.org/.

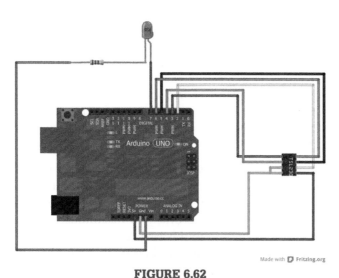

Made with **D** Fritzing.org

FIGURE 6.62

The open-source colorimeter circuit schematic.

Colorimetric methods are used to determine the concentration of dissolved species, relying on the ability of many ionic species to absorb light of one or more specific wavelengths, followingthe Beer–Lambert law shown in Eqn (6.1):

$$A = -\log_{10}\left(\frac{I}{I_0}\right) \tag{6.1}$$

where A is the absorbance (absorbance units), I is the intensity of light passing through unknown, and I_0 is the intensity of light passing through the blank. The absorbance is related to concentration as shown in Eqn (6.2):

$$A = a_\lambda \times b \times c \tag{6.2}$$

where a_λ is the molar absorptivity of the species of interest at a certain wavelength (λ) of light, b is the path length of light through solution, and c is the concentration of the analyte.

The design was evaluated using the closed reflux COD method [47]. The closed reflux colorimetric method used for COD employs potassium dichromate ($K_2Cr_2O_7$) as oxidant, requires only a small sample size, and produces minimal waste. A 2 ml aliquot is added to dichromate solution and allowed to digest at 150 °C for 2 h. Concentration (mg COD/l) is determined by measuring the increase in absorbance at 606 nm, the absorbance peak for the chromic ion, or determining excess dichromate by measuring the absorbance at 440 nm [48,49].

High-range COD digestion vials from Hach Company were used to digest samples produced during unrelated research activities conducted to develop precision and bias statements for a newly developed analytical procedure. In a study [47] comparing the open-source colorimeter and a commercial Hach DR890 portable colorimeter, the open-source colorimeter results (2 mg COD/l) were well within the stated precision of 17 mg COD/l of the Hach method [50]. This indicates that it is suitable for use in any of the standard COD applications.

As usual, the OSH alternative is considerably less costly than closed commercial versions. The open-source colorimeter can be built for one-tenth of the price (approximately US$50) of the least expensive, commercial COD-only instrument available and two orders of magnitude less than the Hach DR890 that it was compared to in this evaluation. Both the hardware design and software are now freely available online and under the creative commons CC-BY-SA license (see Chapter 2 for more details) such that they can be modified and new designs derived. To put this work in further perspective, for about the cost of a standard commercial COD instrument, a research lab can purchase an open-source 3-D printer (or the parts for several—as we saw in Chapter 5) and all the parts to make the open-source colorimeter described here. Thus, perhaps most importantly, the ease and low cost of this approach for developing sensor-based tools enables increasingly sophisticated tools to be used in low-funded developing world laboratories, helping to disseminate open-source appropriate technology for sustainable development [20,51–53]. In addition, high-quality, open-source scientific hardware, such as the colorimeter described here, provides public, nonprofit and nongovernmental institutions, schools and amateur scientists the tools necessary to conduct real science, while driving down the costs of research tools at our most prestigious corporate, government, and academic laboratories [2].

6.4.2. Open-source pH meter

The measure of pH is a measure of the hydrogen ion concentration—or a method to measure the acidity or alkalinity of a solution. By definition, aqueous solutions at 25 °C with a pH less than seven are acidic, while those with a pH greater than seven are basic or alkaline. A pH level of 7.0 at 25 °C is defined as "neutral" because the concentration of H_3O^+ equals the concentration of OH^- in pure water (H_2O). A *pH meter* is an electronic device used for measuring the pH of a liquid. A typical pH meter consists of a measuring probe (a glass electrode, which is a type of ion-selective electrode made of a doped glass membrane that is sensitive to a specific ion) connected to an electronic meter that measures and displays the pH reading. pH measurements are important in a long list of applications including environmental science (e.g. environmental monitoring of soils, water of rivers, lakes, and rain), medicine, biology, chemistry, agriculture, aquaculture, forestry, food science, oceanography, civil engineering (e.g. monitoring of sewage treatment tanks), chemical engineering, nutrition, water treatment and water purification, and many other applications. It is a useful and fundamental property needed in dozens of fields.

The open-source pH meter discussed in this section was developed by Carlos Neves, a Brazilian scientist with a dedication to OSH.[60] The open-source pH meter is a glass electrode pH meter using the Arduino microcontroller detailed in Chapter 4 and is thus compatible with the Freeduino.[61] Full details (hardware, software, operation manual, etc.) can be downloaded for free[62] with the code under a GNU GPL v2 and the content under CC-BY-SA 3.0.

Similar to the open-source environmental chamber in Chapter 4 and the open-source colorimeter in the previous section, the pHduino operates as a stand-alone unit using an LCD to display the pH and the temperature data like any commercial pH meter. Again, similarly, you can control it using a computer by USB port in addition to loading up the firmware. For this design, the signal gain (slope) and the signal offset are adjusted manually by trimpots and the signal is compensated by a temperature sensor. Unlike other examples in this book, this particular design can be more expensive than the low-end commercial pH meter bench instruments that are mass produced due to the tool's vast collection of applications. However, the pHduino has significant advantages, which provide it with superior value. It is interfaceable, programmable, expandable, and, of course open and free, so you are free to adapt it to your specific application without paying additional money to a vendor. As it develops and matures, similar to other areas of complex open-source scientific tools, making and, specifically, printing it from functional materials should reduce costs below those that can be manufactured conventionally and sold.

The printed circuit board (PCB) layout is shown in Figure 6.63, the electronic schematic is shown in Figure 6.64, and the Arduino shield details are shown in Figure 6.65. The completed pHduino is shown in Figure 6.66(a) with details in Figure 6.66(b).

---

[60]You can follow updates to the pHduino development on Carlos's blog. http://phduino.blogspot.com/.
[61]Freeduino is a collaborative open-source project to replicate and publish Arduino-compatible hardware files. The Freeduino Eagle SCH, BRD and Gerber production files allow users to create boards that are 100% functionally, electrically and physically compatible with Arduino hardware. While Arduino is a protected trademark, Freeduino comes with a free and unrestricted license to use the Freeduino name, available for any use. This means you can do whatever you want with their files. The idea here is to make available the Eagle files you would need to make your own Freeduino variant board. So, for example, you could make a simpler board specifically for the open-source colorimeter or the pH meter in this chapter. For more information, see http://www.freeduino.org/about.html.
[62]Freeduino designs http://www.freeduino.org/freeduino_open_designs.html.

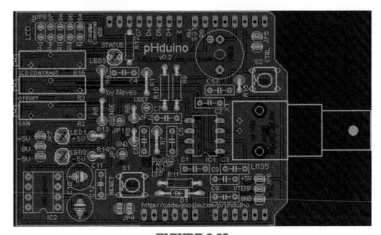

FIGURE 6.63

The printed circuit board (PCB) layout of the pHduino.

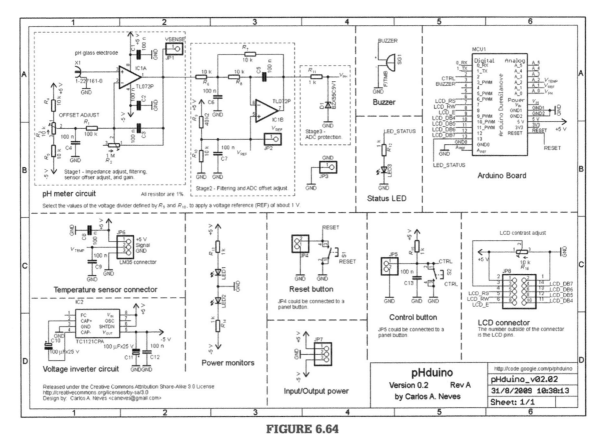

FIGURE 6.64

The electronic schematic for the pHduino.

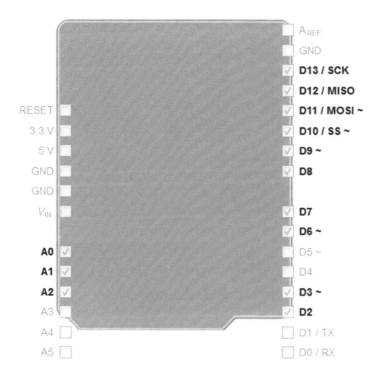

Note:
A0: pH analog signal.
A1: voltage reference.
A2: temperature analog signal.
D2: control button.
D3: buzzer.
D6-D12: LCD.
D13: status LED.

FIGURE 6.65

The Arduino shield details for the pHduino.

In addition, the pHduino can be improved with the addition of an open-source (hardware and software) pHduino Datalogger[63] shown in Figure 6.67. The datalogger is a shield using an I2C real-time clock (RTC) and an I2C EEPROM memory. The circuit is relatively simple as can be seen in the PCB layout for the datalogger in Figure 6.68 and the electronic schematic shown in Figure 6.69. Users can mount the datalogger using a breadboard or a universal prototyping board. The datalogger is functional, but somewhat limited as the AT24C512 I2C EEPROM (512 kB) is not a large amount of memory, although

---

[63]pHduino Datalogger http://code.google.com/p/phduino/wiki/pHduinoDatalogger.

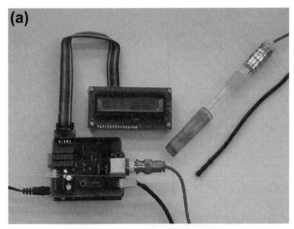

FIGURE 6.66

(a) The completed pHduino and (b) details.

it is inexpensive and easy to use. The DS1307 I2C real-time clock is a standard RTC for microcontrollers. Finally, a CR2032 battery is used to retain the parameters. For full build instructions and details, see Carlos' detailed and updated web pages.

You are encouraged to build on these open designs and improve the resolution of this instrument to compete with the most expensive and sophisticated commercial pH meters. The pHduino was designed to be a simple, easy to understand, and easy to modify analytical tool like the open-source spectrometers to be discussed in Section 6.6. Like the open-source colorimeter, the pHduino could also be immediately improved to be a field instrument (beyond a bench top!) by providing a case for it similar to the Geiger counter in the previous

FIGURE 6.67
Open-source pHduino datalogger.

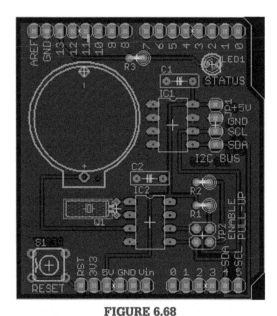

FIGURE 6.68
PCB layout for the pHduino datalogger.

section. Again, we are at the beginning and just scratching the surface of open-source scientific hardware. You should expect more sophisticated high-performance tools to be on the Internet shortly—may be you can help accelerate their development?

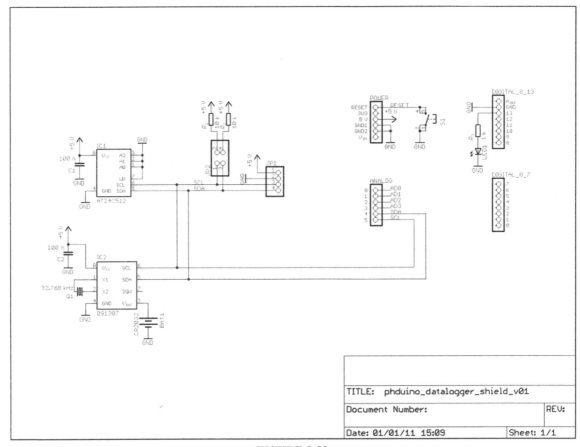

FIGURE 6.69
The electronic schematic for the pHduino datalogger.

6.5.   BIOLOGY: OPENPCR, OPEN-SOURCE CENTRIFUGES AND MORE

OSH can also be used for biology, biological science, and bioengineering. In this section, we will look at designs for an open-source polymerase chain reaction (PCR) machine, open-source centrifuges, and many other biological research tools. The results of the equipment from the designs are discussed similarly to the last section to provide data for conclusions about future development of other open-source analytical tools discussed in Chapter 7.

6.5.1.   OpenPCR

PCR is a biochemical technology used in molecular biology research to amplify a relative handful of copies of a piece of DNA (or even a single piece) by making as many as you like (e.g. thousands to millions of copies). The replications of a particular DNA sequence is useful for literally dozens of applications in biology and medical science, which include (1) DNA cloning for sequencing; (2) DNA-based phylogeny, or functional analysis of genes and analyzing gene expression levels; (3) the detection and diagnosis of infectious diseases including both viral and bacterial infections (thus it is also useful for doing water quality testing or food safety testing); (4) diagnosis of hereditary diseases; and (5) the identification of genetic fingerprints, which is useful in forensic science. For all its utility, the PCR method is actually fairly simple; it uses cycles of repeated heating and cooling (e.g. thermal cycling) of the reaction for DNA melting and enzymatic replication of the DNA. Short DNA fragments called primers, which contain sequences complementary to the target region along with a DNA polymerase, enable selective and repeated amplification. A chain reaction is established where the DNA template is exponentially amplified because the DNA generated is itself used as a template for further replication. The more you have, the more you get and the faster you can make the next batch. The method is extremely effective, although the commercial equipment is staggeringly expensive, and this has set up a large and ugly intellectual property battle. The IP battle continues to this day, despite the fact that the original PCR and Taq polymerase patents expired in 2005. For more details on negative effects of intellectual property on scientific progress and innovation, see Chapter 3. The answer to the unnecessarily complicated world of intellectual property when it comes to science is, of course, to develop open-source PCR equipment.

Fortunately, this has already been accomplished as you can see in Figures 6.70(a) and (b), which show the OpenPCR in action. OpenPCR is a low-cost, yet accurate, thermocycler you build yourself, capable of reliably controlling PCR reactions for DNA detection, sequencing and other applications.[64] You can buy the OpenPCR kit for $600 and build it in about 3 h using only a Phillips head screwdriver, a 2 mm flathead screwdriver and a pair of pliers. The assembly is reminiscent of the old laser-cut wood 3-D printers—only much easier and less time-consuming to assemble. OpenPCR also utilizes the Arduino platform as described in detail in Chapter 4. The OpenPCR is completely open source including the Arduino software,[65] CAD files,[66] the PCB design,[67] and the BOM.[68]

The OpenPCR has a sample capacity of 16 0.2 ml tubes. It can operate at temperatures from 10 to 100 °C (±0.5 °C). The average ramp rate is quite fast and

---

[64]Open PCR http://openpcr.org/.
[65]OpenPCR software http://github.com/jperfetto/OpenPCR.
[66]OpenPCR CAD files http://download.openpcr.org/design/OpenPCR-CAD-1.0.zip.
[67]OpenPCR board http://download.openpcr.org/design/OpenPCR-BoardDesign-1.0.zip.
[68]Open PCR BOM https://docs.google.com/spreadsheet/ccc?key=0AiaWH3PL_9CndG1GbmtSZ1I3aGd kQkE4YVhwZ3o5c1E#gid=.

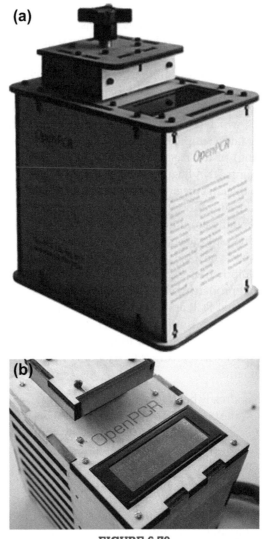

FIGURE 6.70
(a) The OpenPCR case and (b) running detail.

is about a degree per second. The heat block for the OpenPCR is currently machined out of a highly thermally conductive alloy of aluminum, until 3-D printing of metals becomes common.[69] The block is insulated at the sides to ensure a uniform temperature across the block and tight contact with standardized 200 μl PCR tubes. This results in a temperature which is uniform within

---

[69]That may be much faster than commentators have predicted. Our group (along with many others) is planning to build low-cost open-source 3-D printers for metal in the short term.

±0.3 °C. The OpenPCR is completely programmable/automatable—so you can set up the initial, cyclic and final steps from either a connected computer or after it has been set up to run as a stand-alone unit. One of the primary features of the OpenPCR is the heated lid that can range from the ambient temperature in your lab up to 120 °C. The heated lid heats up the tops of your tubes and ensures that condensation is prevented, thereby improving your results by ensuring that your reagent concentrations remain correct throughout the cyclical process—just as high-end PCRs do. If this was a closed, commercialized PCR and you were unhappy with the features, you could call and complain to the company, go shopping again, or start from scratch and build one yourself. That is not a very good range of options. However, as with all OSH tools, with the OpenPCR, you can change or enhance it to meet your specific needs—the "activation energy" to build on the already created open design is much less than that to start from scratch. OpenPCR has already been hacked to do microfluidic PCR, and software has been written for complete laboratory automation.

6.5.2. Open-source centrifuges

A laboratory centrifuge is used to spin liquid samples at high speeds and has many applications in a range of different types of biological laboratories. Centrifuges work using the sedimentation principle, where the centripetal acceleration causes denser substances to separate out along the radial direction (e.g. for particles to collect at the bottom of the tube). There are many types of centrifugation, which are useful in the laboratory including (1) differential centrifugation, which can separate certain organelles from whole cells for further analysis of specific parts of cells; (2) isopycnic centrifugation, which is used to isolate nucleic acids such as DNA; and (3) sucrose gradient centrifugation, which is used to separate cell organelles from crude cellular extracts and to purify enveloped viruses and ribosomes. Microcentrifuges tend to be the most useful in a lab as they are designed for small tubes able to hold from 0.2 to 2.0 ml (micro tubes). In general, these small devices can obtain up to 30,000 g (where g is the acceleration due to gravity at the Earth's surface). Protocols for centrifugation normally specify the amount of acceleration to be applied to the sample, rather than specifying a rotational speed such as revolutions per minute (RPM), which is useful because it enables protocols to span different devices with various rotor lengths. The relative centrifugal force [g] applied to a sample in a centrifuge is given by Eqn (6.3):

$$g = \text{RCF} = 0.00001118 \times r \times s^2 \qquad (6.3)$$

where r is the rotational radius in centimeter and s is the rotational speed in RPM. Here, we will briefly outline several types of open-source centrifuges.

FIGURE 6.71
DremelFuge chuck.

6.5.2.1. DremelFuge

Of all the open-source centrifuge designs, the DremelFuge is perhaps the most elegant. The DremelFuge (as shown in Figure 6.71) is a printable rotor for centrifuging standard microcentrifuge tubes and miniprep columns, which uses a high-speed drill or Dremel tool to spin down samples as shown in Figure 6.72.[70] The Dremelfuge was developed by Cathal Garvey in Ireland to assist in his research with DIYbio.[71]

The primary idea behind the DremelFuge is that it can be used in the field as an extremely inexpensive centrifuge (costing about $50—primarily the cost of the drill—compared to commercial systems starting over $500). It can be used for any application in development needing a microcentrifuge including medical, biochemistry or education in the sciences. It uses industry standard 1.5 ml/2 ml Eppendorf/Microcentrifuge tubes.

The applications of the Dremelfuge are numerous and impressive. For example, when used with a drill at 3000 RPM, the Dremelfuge will deliver over 400 g, enough to comfortably spin down Miniprep samples. When used at 10,000 RPM, on a rotary tool for instance, a Dremelfuge should deliver over 4,400 g, more than enough to spin down bacterial cells. The Dremelfuge has surprisingly competitive technical specifications, when used on high-speed

---

[70]DremelFuge http://www.thingiverse.com/thing:1483.
[71]DIY BIO http://diybio.org/.

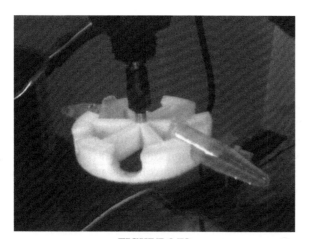

FIGURE 6.72
DremelFuge chuck attached to Dremel tool to spin down samples.

Dremels. For example, at 16,000 RPM, Dremelfuge matches commercial centrifuges. It can do better. If a Dremelfuge is used with a Dremel 300, a maximum speed of 33,000 RPM equates to a force of over 50,000 times earth's gravity, which puts it into so-called "Ultracentrifuge" territory. The latest version (as printed by Shapeways[72]) has successfully spun tubes at this speed.

Obviously, the Dremelfuge is a potentially hazardous tool and significant safety considerations and common sense are necessary when using it.[73]

The general steps for making and using a Dremelfuge are the following:

1. Print the Dremelfuge chuck on a RepRap or buy Dremelfuge preprinted from a 3-D printer supplier. When printing, use the maximum infill for strength and stability.
2. Attach the Dremelfuge chuck securely, either by tightening a chuck securely or screwing a rotary tool disc holder securely to the center of the Dremelfuge.
3. Seat your drill or rotary tool so that Dremelfuge's shaft/axle is vertically oriented. You can do this using conventional building materials, OpenBeam (as was discussed in the beginning of this Chapter) or Makerbeam, or even conventional chemistry setups. Then, place the drill/

---

[72]For those of you that have not put together a RepRap following the directions in Chapter 5, you can still get your hands on 3-D printed objects using a commercial printing service like Shapeways. Here is the link for the DremelFuge http://shapeways.com/shops/labsfromfabs.
[73]Centrifuges require some care in order to be used safely, and a printed centrifuge in particular will have hazards that you should be ready for if you try to use it. Consult your laboratory's safety officer before using this device and always wear your safety glasses.

tool with the Dremelfuge into a metal chamber (e.g. a metal cooking pot) for safety. Wear eye gear and other appropriate personal protective items to protect yourself in case of disintegration.

4. Start at the lowest speed and ramp up the RPMs. Do this first with no tubes or loads attached and test the Dremelfuge for safety at the speeds you intend to use it.

5. Once proven safe at the intended speed, you can start to test and use Dremelfuge under load, that is with desired size of the lab samples and perhaps water. Once this is done, move to the lab samples. Make certain at all times that identical tubes or columns are used, with identical amounts of fluid or mass on either side to maintain balance. *Always balance the Dremelfuge or accidents may result.*

6.5.2.2. *Microcentrifuge*

If you are willing to trade a little performance for a very low-cost centrifuge, consider the table-top microcentrifuge[74] (shown in Figure 6.73) developed and open-sourced by Thingiverse user Tinytim.

The BOM includes the following:

1 × Graupner speed 400 motor (or similar sized motor, available on the Internet or your local model shop)
1 × 220 V switch
1 × 220 V to 6 V trafo (Shown: a small print-trafo 6 V, 3.2 VA, size 35 × 42 × 30 mm)
1 × bridge rectifier
1 × pot suitable to regulate your motor speed (the size depends on your motor's Ohm)
2 × 3 mm screws (about 5 mm long)
2 × 3 mm threaded rod (about 1 cm long), with a slit sawed into it as an inexpensive replacement for a headless screw
2 × 3 mm nut
An electric cord
Printed parts (printed in PLA, 1.2 mm wall-size, 50% infill for all parts—0.2 mm slicing using Cura).

Print your parts on a RepRap (built in the last chapter) or similar. Both the rotor and casing bottom can be printed without support material; however, the casing top does demand support printing.[75] Next, wire and assemble. For your first test, it is recommended that you take it up to speed inside a steel

---

[74]Microcentrifuge http://www.thingiverse.com/thing:33818.
[75]This can also be accomplished printing with a 2-headed RepRap and using PVA as a support printing material.

FIGURE 6.73
Table top microcentrifuge.

pot, similar to the protocol for the DremelFuge. Afterwards it can function as a desktop centrifuge. If you are concerned about safety or your own making skills, you can build or print a safety shield for it as shown in the next example. The example shown here can spin up to 16,400 RPM at 7.2 V, but was only run at 6 V. Again, be careful particularly with the printing as a rotor that is not printed carefully may break during centrifugation and cause damage or injury.

6.5.2.3. USB powered cytocentrifuge

The third type of previously designed centrifuge is the USB powered cytocentrifuge[76] developed by Thingiverse user siderits (Dr Richard Siderits of Robert Wood Johnson University Medical School) for under $10. It uses a small DC motor (1.5–6 V, 4000 RPM) and a DC motor speed controller kit to create the USB powered biocentrifuge shown in Figure 6.74. Although you can use longer rotor arms, the test centrifuge has a radius of 40 mm. The motor will spin up to 4000 RPM on 4.5 V/100 mA USB port power giving a centrifugal force of 500 g. The USB centrifuge can be placed in either a printable enclosure or any other appropriate enclosure to make a functional very inexpensive low-end centrifuge.

To fabricate the USB centrifuge, print out the parts, purchase an appropriate DC motor, the speed controller and assemble. Similar to the Dremelfuge, the printed parts that experience physical forces should be printed at 100% fill. In the example shown, the US$8 DC motor speed control PWM is part number FK804 from Bakatronics. The potentiometer controls the width of the pulses to control the

---

[76]USB powered cytocentrifuge http://www.thingiverse.com/thing:46448.

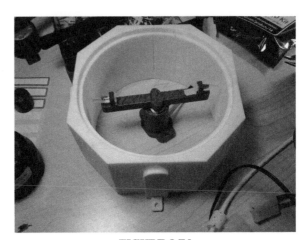

FIGURE 6.74
USB powered cytocentrifuge.

speed of the motor. It can be mounted directly on the unit, remotely on a control panel or in a printed enclosure, or in FB enclosure to make a hand-held throttle. The USB centrifuge uses a 12C power supply, but can be used as low as five VDC. The housing is FB03 and the PCB is 1.8″ × 1.3″. The maximum current is 1.5 A. In addition, you can add a DPDT slide switch for reversing the direction of the motor. If you do not have a small DC motor that has speed up to 4000–5000 at 1.5–6 V, you can salvage, you can buy one for about $3 from numerous Internet vendors. Lastly you will need an old USB cable.

Dr Richard Siderits is an excellent example of the open-source scientist of the future. He is an Anatomic and Clinical Pathologist, fellowship trained in Experimental Pathology and he is currently an Associate Professor at Robert Wood Johnson University Medical School, Department of Pathology and Laboratory Science. In addition to his research, he especially enjoys teaching rapid prototyping and science, technology, engineering and mathematics (STEM) principles (including, of course, 3-D Printing) as they apply to the medical sciences and the history of medicine. The shear volume of his contributions to open-source science is already substantial. For a full list, see his Thingiverse user page.[77] In addition to the cytocentrifuge, he has shared designs for many useful tools such as 3-D printable rapid fluid filters[78] as seen in Figure 6.75, PCR tube racks[79] (Figure 6.76), cassette racks[80] (used to organize cassettes that

[77]Dr. Richard Siderits' designs http://www.thingiverse.com/siderits/designs.
[78]3-D printable rapid fluid filters http://www.thingiverse.com/thing:27933.
[79]PCR tube racks http://www.thingiverse.com/thing:25450.
[80]Cassette rack http://www.thingiverse.com/thing:28575.

FIGURE 6.75
3-D printable rapid fluid filters.

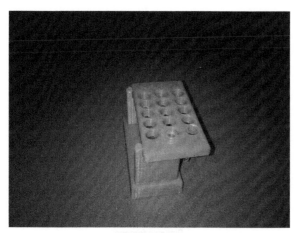

FIGURE 6.76
PCR tube racks.

hold tissue samples and large specimens in surgical pathology and shown in Figure 6.77), and platypus forceps[81] (Figure 6.78), which hold a portion of tissue so that a scalpel or microtome blade can slide in between and section the tissue at a uniform thickness.

Currently, he is working on several projects that should be coming out as this book is printed, including a home bioprinter made using Makerbeam[82]

---

[81]Platypus Forcep http://www.thingiverse.com/thing:31868.
[82]Makerbeam: http://www.makerbeam.eu/.

FIGURE 6.77
3-D printable cassette racks.

FIGURE 6.78
Platypus forceps.

connectors (which he has already developed as seen in Figure 6.79[83]) and step-per motors, an auto slide stainer for histology (two-slide version), an auto immunohistochemical stainer (single-slide version) and a USB-powered real-time quantitative PCR module. Finally, if you want to see how science used to be done or want to better explain it to your students, he is developing a line of working 3-D printed reproductions of vintage laboratory equipment circa 1850–1920.

---

[83]Connectors for 20 mm MakerBeam http://www.thingiverse.com/thing:28179.

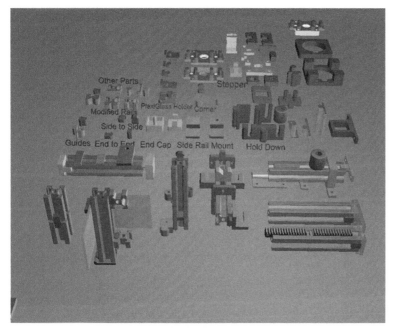

FIGURE 6.79
Connectors for 20 mm MakerBeams.

6.5.3. Other useful open-source biological research tools

There are many other useful open-source biological research tools already available and this book does not even attempt to catalog all of them, but I would like to pique your interest in using and sharing OSH, so consider the following simple examples of tools that can save your biolaboratory money.

Racks for vials and test tubes proliferate on the open-source scientific hardware community. For example, you can get test tube racks that are made with laser cutters[84] as designed by Thingiverse user Eithen and shown in Figure 6.80 or with 3-D printers[85] as designed by Thingiverse user _Kid_ as shown in Figure 6.81.

As complexity is free with the sharing of digital designs, it is just as easy to replicate an inexpensive test tube rack as it is to make an $850 magnetic rack. For example, Andrew Adey (Thingiverse user Acadey) is a graduate student in the Shendure Lab, University of Washington, Genome Sciences who became frustrated with the seemingly criminal charges for magnet racks.[86] He designed a

---

[84]Laser cut test tube racks http://www.thingiverse.com/thing:22728.

[85]3-D Printable test tube racks http://www.thingiverse.com/thing:69911.

[86]Acadey Designs http://www.thingiverse.com/acadey/designs.

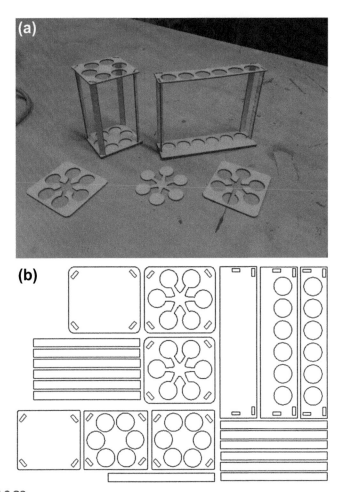

FIGURE 6.80

Laser cut test tube racks: (a) the cut and assembled rack and (b) the pattern for laser cutting.

96-well plate/0.2 ml strip tube magnet rack[87] shown in Figure 6.82. It can per-form magnetic bead separation for a 96-well plate, or alternatively, a 96-well plate with the wells cut out can be taped on top to hold eight tube strips. The magnets are available for about $6 a piece[88], making it possible to justify the cost of a RepRap with a single (or at most two) standard commercial magnetic racks that normally run between $450 and $900. Similarly, he designed a magnetic rack for eight 1.5 ml tubes as shown in Figure 6.83, which can be put together with 70 cent magnets, saving you hundreds of dollars per rack.[89] As we have

---

[87]96-well plate/0.2 ml strip tube magnet rack http://www.thingiverse.com/thing:79430.
[88]K&J Magnetics http://www.kjmagnetics.com/proddetail.asp?prod=BZ082.
[89]Magnetic Rack for 8, 1.5 ml tubes http://www.thingiverse.com/thing:79424.

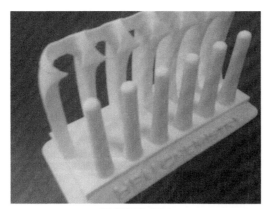

FIGURE 6.81
3-D printable test tube racks.

FIGURE 6.82
96-well plate/0.2 ml strip tube magnet rack.

seen throughout this book, open designs are iterative and are able to be easily built upon. The University of Washington plans on climbing the ladder of open-source sophistication and are obtaining a high-resolution printer for more fine-scale designs.[90] They hope to build some machines for automated sample processing using 3-D printed items along with fluidics controlled by the open-source Raspberry Pis and Arduinos. You may want to consider helping them.

---

[90]It should be noted that the RepRap community is also pushing the resolution of low-cost 3-D printers down, as well as diversifying the printable feedstock to all of the material families (e.g. ceramics and metals).

FIGURE 6.83
A magnetic rack for eight 1.5 ml tubes.

There are a growing number of organizations devoted to producing OSH for science. One such group is Hackteria.org,[91] which was designed as a community platform to encourage the collaboration of scientists, hackers and artists to combine their expertises, share instructions on how to work with life science technologies, and cooperate on the organization of workshops, temporary labs, hack-sprints and meetings. It was started in 2009 by Andy Gracie, Marc Dusseiller and Yashas Shetty. What has developed is a rich wiki-based web resource for people interested in or developing projects that involve bioart, open-source software/hardware, DIY biology, art/science collaborations and electronic experimentation. For a feel of how Hackteria combines art and science, consider the Tardi-GelBox for electrophoresis designed by Thingiverse user dusjagr shown in Figure 6.84.[92] The box is an open-source project that is laser cut and sports a tardigrade on it. Why a tardigrade? They are pretty cool,[93] but I am not exactly sure. That perhaps is the point that Hackteria strives to make. With the collaborative open-source paradigm, not only can less-expensive high-performance tools be built and shared with the world, but at the same time, we can make them customized works of art to increase the aesthetic appeal, ability for our

---

[91]Hackteria http://hackteria.org/.

[92]Tardi-GelBox http://www.thingiverse.com/thing:71638.

[93]These water-dwelling 1-mm-long eight-legged creatures sometimes called water bears are one of the most complex of all known polyextremophiles that can survive in horrendous conditions. For example, tardigrades can withstand temperatures from just above absolute zero to well above the boiling point of water, as well as extreme pressures, solar radiation, gamma radiation, ionic radiation and have even lived through the vacuum of space. A lot of people would consider this pretty cool. May be you do not. That is ok because you now have the ability to turn all your scientific equipment into works of art with your own favorite creatures.

FIGURE 6.84
Tardi-GelBox.

students to express themselves or just the general excitement for research in our laboratories. The Tardi-GelBox is laser cut in acrylic and made water tight with acryl-glue. After assembly, you simply cast your agar-gel, pour the running buffer (e.g. salt water), add your DNA samples, plug into a DC power supply (15–100 V) and enjoy. Hackteria has over 70 other low-cost DIY designs such as microscopes, microfluidics, incubators, spectrometers, bioprinting and more that are best accessed through their wiki.

6.6. CHEMISTRY: SPECTROMETERS AND OTHER CHEMICAL RESEARCH TOOLS

OSH can also be used for chemistry and chemical engineering. In addition to the work of Symes et al. [15], which has demonstrated using a 3-D printer as reactionware for chemical synthesis in a way that makes open-source digital chemistry a reality, the open-source paradigm can make more conventional (yet less-expensive and customizable) chemical research equipment. In this section, we will look at designs for spectrometers and other research tools useful to those skilled in the chemical sciences. The results of the chemical equipment from the designs are discussed similarly to the last section to provide data for conclusions about future development of other open-source analytical tools discussed in Chapter 7.

6.6.1. Open-source spectrometers

One of the best examples of the OSH approach being applied to research equipment while also contributing to STEM outreach to bring more young people into the sciences is being performed by The Public Laboratory for Open

Technology and Science (also known as Public Laboratory).[94] The Public Laboratory mission is "Using inexpensive DIY techniques, we seek to change how people see the world in environmental, social, and political terms. We are activists, educators, technologists, and community organizers interested in new ways to promote action, intervention, and awareness through a participatory research model." Thus, using inexpensive components and open-source software, Public Laboratory has developed an Android-based cell phone spectrometer with a range of 400–900 nm and a resolution of 3 nm [54]. The properties of the Public Laboratory spectrometer are comparable to those of a commercial mobile hand-held visible light spectrometer that costs over $2500.[95] A screenshot of a cell phone equipped with the Public Laboratory Spectral Workbench software is shown in Figure 6.85.

Public Laboratory maintains a detailed site for teaching everyone about spectroscopy. It also provides comprehensive instructions on how to build several varieties of spectrometers including (1) a foldable mini-spectrometer, (2) a smartphone spectrometer and (3) a desktop spectrometry kit along with all the necessary open-source software to analyze scans.[96] The initial versions of the Public Laboratory spectrometers were made by crudely taping a small black paper case and a slice of DVD-R to the back of my Android[97] phone. This worked, but they have progressed beyond these initial designs. Now using the powerful tool of open-source 3-D printing (discussed in Chapter 5 in detail), you can print a superior version of the Public Laboratory spectrometer (v3.0)[98] as seen in Figure 6.86, which was released under an open-source CERN OHL v1.1 license.

Public Laboratory ran a successful crowd-funding campaign for their spectroscopy project and now they offer a desktop spectrometry kit for $40. This again represents more than an order of magnitude decrease in spectroscopy costs for those already in possession of a computer or an appropriately endowed smartphone. This spectrometer greatly expands the potential users of spectroscopy. It again can be used to encourage interest in science and engineering in young people. For example, it can be used to identify dyes in laundry detergent, to test grow lamps, to probe water quality and to analyze food. Public Laboratory writes detailed protocols for each application, with the hope of making science more accessible and a part of everyday life [54].

---

[94]Public Laboratory http://publiclaboratory.org/home.
[95]Public Laboratory Visible Light Spectrometer http://publiclaboratory.org/notes/warren/11-57-2011/visible-light-spectrometer-2660-4-7nm-bandwidth.
[96]Public Laboratory Spectrometer http://publiclaboratory.org/tool/spectrometer.
[97]Android is also an open-source Linux-based operating system (provided by Google under the Apache License) designed primarily for smartphones and tablets and touchscreen mobile devices. For full details, see http://source.android.com/.
[98]Public Laboratory spectrometer (v3.0) http://www.thingiverse.com/thing:49934.

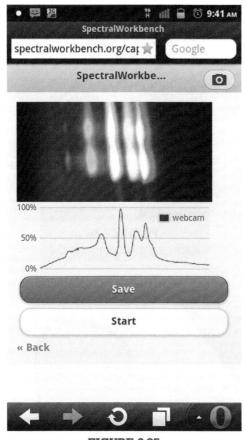

FIGURE 6.85

A screenshot of a cell phone equipped with the Public Laboratory Spectral Workbench software.

Public Laboratory is not alone in investigating open-source spectroscopy. Myspectral has developed the Spectruino,[99] which uses the Arduino platform (discussed in Chapter 4) to provide a UV/Vis/IR spectrometer, which measures light intensity as a function of wavelength with a combination of diffraction grating and a linear CCD camera. The Spectruino, which is partially open source, is shown in Figure 6.87.

The Spectruino is more robustly built as it consists of a stainless steel and aluminum case. However, the design still suffers from the detector being fixed to the case with a material that is affected by temperature, which limits its use for some high-precision applications. This, of course, is only the beginning of open-source

---

[99]Spectruino http://myspectral.com/#technical.

FIGURE 6.86
Public Laboratory spectrometer v3.0.

FIGURE 6.87
The Spectruino.

spectroscopy as the projects from Myspectral and Public Laboratory are growing and improving as more scientists add to the designs and software. For fans of Star Trek, it is pretty clear where we are going—and in fact we are already at the beginning stages of an open-source tricorder.[100] Many of the Arduino-based scientific tools that have already been developed can be integrated together, so

---

[100]In the Star Trek series and movies, a tricorder is a multifunction hand-held device used for sensor scanning, data analysis, and recording data. There are all different flavors of tricorders—such as those for engineering, planet exploration, and medical diagnosis.

FIGURE 6.88
The Science Tricorder Mark 2 prototype sensor board.

as each group or team of specialists continue to work on their own piece of the scientific puzzle, once it becomes developed enough, it will be possible to bring them all come together to form a tool that really is right out of the science fiction books. A very ambitious open-source Tricorder Project is now well into development.[101] The Science Tricorder Mark 2 prototype sensor board shown in Figure 6.88 contains 10 different sensing modalities, organized into three main themes: atmospheric sensors (temperature, humidity, and pressure), electromagnetic sensors (magnetic flux, color, an IR thermometer, and ambient light levels), and spatial sensors (GPS location, ultrasonic distance sensor, accelerometer and a gyroscope).[102] Again, this is only scratching the surface of what we can now do with sensor technology and there are many organizations pushing the technology forward.[103] It is clear we will see more increasingly sophisticated open-source spectrometry projects in the near future.

The open-source spectrometers significantly reduce the costs associated with many experiments that need them, while also enabling the production of highly customizable designs for specialized experiments. These reductions in costs and increases in flexibility are likely to boost technical development in any field that utilize spectrometers. At the same time that it assists the research community, it can also radically reduce the cost of education, amateur and DIY science [2]. This is thus likely to encourage more DIY scientists and young people to enter STEM fields [55,56]. These developments provide interesting possibilities in the future, where it is now quite possible to consider the

---

[101]The Tricorder Project http://www.tricorderproject.org/.
[102]The Tricorder Project Mark 2 http://www.tricorderproject.org/tricorder-mark2.html.
[103]For example, there is the Tricorder X Prize, a $10 million competition to develop a hand-held portable, wireless device that monitors and diagnoses your health conditions to allow unprecedented access to personal health metrics. http://www.qualcommtricorderxprize.org/.

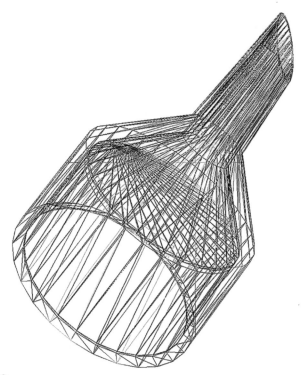

FIGURE 6.89
An open-source printable Buckner funnel shown in MeshLab.

general public performing their own real-time chemical analyses with their cell phones. Similar to the crowd-sourced data gathering with the Arduino-based Geiger counters, the public could provide highly reliable geographically tagged data on air, water and food quality, etc.

6.6.2.   3-D printable chemical equipment

In the last section, we looked at various test tube racks and stands, which would be useful to chemists in addition to biologists, but there are also many other both simple and more advanced (and costly) 3-D printable components for chemistry laboratories. For example, consider the Buckner funnel and open-source printable example,[104] of which is shown in MeshLab[105] (open-source 3-D mesh processing software) Figure 6.89.

Buchner funnels are relatively common as they are used for vacuum-assisted filtration in all manner of experiments. On top of the funnel-shaped part, there

---

[104]Buckner funnel http://www.thingiverse.com/thing:25188.
[105]Meshlab http://meshlab.sourceforge.net/.

FIGURE 6.90

A pipette stand made from printed parts and nonrefillable HPLC columns.

is a cylinder with a perforated plate separating it from the funnel. The designer (Thingiverse user: chowderhead) made the OpenSCAD script parametric, so you can have any size of Buchner funnel you would like. All the dimensions of the Buckner funnel are based on the diameter of the filter paper you would like to use (as filter media comes in a variety of diameters). Because of the limitations of printing on most single-material 3-D printers, in this design, the necessary perforations are not included and thus need to be drilled afterward. However, in the near future, as many open-source 3-D printers are moving to two or more print heads, one of which can print a sacrificial support material like sugar or polyvinyl alcohol (PVA), I am confident that someone will design a one-step printable Buckner funnel. For the plate that needs drilled, it will need 100% infill, but if you are printing the holes, you can get away without using as much plastic. Also note that depending on the quality and fine tuning of your printer, you may have relatively porous walls. To eliminate this, as it would ruin the functionality of the funnel, you will need to seal it and depending on your printing material, there are several options. If you printed it in ABS, you can use a mixture of acetone and ABS (e.g. bad prints) to paint your funnel. Likewise PLA can be smoothed with a dip treatment in dichloromethane (CH_2Cl_2 or DCM).

It is also possible to take advantage of discarded equipment with the help of a laboratory 3-D printer to make other items you need around the lab. For example, consider the pipette stand[106] designed by thingiverse user: M_Sanna shown in Figure 6.90.

---

[106]Pipette stand http://www.thingiverse.com/thing:91209.

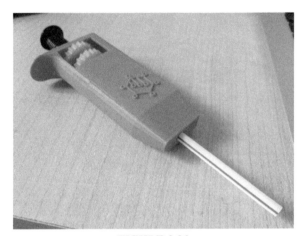

FIGURE 6.91
An open-source printable adjustable volume pipette.

The ability to fabricate custom connectors makes it possible to use nonrefillable high-performance liquid chromatography (HPLC) columns as a constructor set to make the pipette rack. Pipette racks are probably not a major expense in your lab and usually cost $85–140, but again this is money perhaps better spent elsewhere. You do not need to stop there. You can make your own adjustable volume pipette[107] as shown in Figure 6.91, which was developed by Dr Konrad Walus of the Department of Electrical and Computer Engineering at the University of British Columbia. He is well known in the maker community for his clever low-cost 3-D printable scientific designs for hobbyists (e.g. 3-D printable microscopes), but the broader aim of his research group is to develop printing technology that is capable of producing active electronic, mechanical, and chemically sensitive devices. This is the next stage in open-source 3-D printing as more complex, active, smart and functional materials are printed into objects turning them into sophisticated devices. In addition, his lab has already begun to look at biologically useful structures (e.g. programmatically defined tissues by 3-D printing multicellular assemblies) in new variations of the basic printing technology. His group is specifically targeting the fabrication of 3-D tissue constructs for use in drug screening and testing. The potential for enormous impact and radical improving while reducing the costs of modern medicine are staggering. Perhaps a final example shows how the Walus group is repaid for sharing. Following the open-source model, Walus's adjustable volume pipette has already been simplified and improved and reshared[108] as seen in the new design show in Figure 6.92 developed by Thingiverse user: aliekens.

[107]Adjustable volume pipette http://www.thingiverse.com/thing:64977.
[108]Adjustable Volume Straw Pipette (simpler plunger) http://www.thingiverse.com/thing:74502.

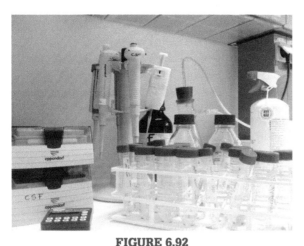

FIGURE 6.92

An open-source printable simplified adjustable volume pipette.

This version of Walus's pipette has bulkier top components and is easier to print than the original, which extends the designs to a larger community of users. Aliekens' primary modification was to remove the threaded bolt from the plunger in Walus's design, and redesign the plunger so that a 75 mm M4 bolt could be inserted through the length of the plunger replacing the original thread. Because of this change, the thumb gears were also modified to accept the M4 rod. To ensure a closed surface, you should seal the membrane chamber with a few coats of ABS glue (ABS scraps mixed with acetone). In addition, this same ABS + acetone mixture should be used to fit the straw where the straw enters the seals, which eliminates any possibility of air escaping. Aliekens has successfully used a double membrane made of lab gloves with 200 µl tips to obtain below 10 µl consistency. This potentially replaces the pipettes that can cost several hundred dollars with a few dollars of materials and it will only get better. The future of open-source chemistry equipment is looking bright.

REFERENCES

[1] Jones R, Haufe P, Sells E, Iravani P, Olliver V, Palmer C, et al. RepRap – the replicating rapid prototyper. Robotica 2011;29:177–91.

[2] Pearce JM. Building research equipment with free, open-source hardware. Science 2012; 337:1303–4.

[3] Sells E, Bailard S, Smith Z, Bowyer A, Olliver V. RepRap: the replicating rapid prototyper-maximizing customizability by breeding the means of production. In: Pillar FT, Tseng MM, editors. Handbook of research in mass customization and personalization. Strategies and concepts, vol. 1. New Jersey: World Scientific; 2009. p. 568–80.

[4] Dimiduk TG, Kosheleva EA, Kaz D, McGorty R, Gardel EJ, Manoharan VN, et al. A simple, inexpensive holographic microscope. Digital holography and three-dimensional imaging. Osa technical digest (CD). Opt Soc Am 2010:JMA38.

[5] Sun R, Bouchard MB, Burgess SA, Radosevich AJ, Hillman EM. A low-cost, portable system for high-speed multispectral optical imaging. Biomedical optics. Osa technical digest (CD). Opt Soc Am 2010:BTuD41.

[6] Teikari P, Najjar RP, Malkki H, Knoblauch K, Dumortier D, Gronfier C, et al. An inexpensive Arduino-based LED stimulator system for vision research. J Neurosci Methods 2012;211:227–36.

[7] Zhang C, Anzalone NC, Faria RP, Pearce JM. Open-source 3D-printable optics equipment. PLoS One 2013;8(3):e59840. http://dx.doi.org/10.1371/journal.pone.0059840.

[8] Jones RO, Iravani P, Bowyer A. Rapid manufacturing of functional engineering components. Bath, UK: University of Bath; 2012.

[9] Baechler C, DeVuono M, Pearce JM. Distributed recycling of waste polymer into RepRap feedstock. Rapid Prototyping J 2013;19:118–25.

[10] Kreiger M, Pearce JM. Environmental impacts of distributed manufacturing from 3-D printing of polymer components and products. MRS Online Proc Libr 2013;1492; mrsf12-1492-g01-02.

[11] Pearce JM, Blair CM, Laciak KJ, Andrews R, Nosrat A, Zelenika-Zovko I. 3-D printing of open source appropriate technologies for self-directed sustainable development. J Sustainable Dev 2010;3:17.

[12] Kreiger M, Anzalone GC, Mulder ML, Glover A, M Pearce J. Distributed recycling of postconsumer plastic waste in rural areas. MRS Online Proc Libr 2013;1492; mrsf12-1492-g04-06.

[13] Pearce JM. Make nanotechnology research open-source. Nature 2012;491:519–21.

[14] Pearce JM. Open-source nanotechnology: solutions to a modern intellectual property tragedy. Nano Today. Available online 2013 http://dx.doi.org/10.1016/j.nantod.2013.04.001.

[15] Symes MD, Kitson PJ, Yan J, Richmond CJ, Cooper GJT, Bowman RW, et al. Integrated 3D-printed reactionware for chemical synthesis and analysis. Nat Chem 2012;4:349–54.

[16] Cardona A, Saalfeld S, Schindelin J, Arganda-Carreras I, Preibisch S, Longair M, et al. TrakEM2 software for neural circuit reconstruction. PLoS One 2012;7:e38011.

[17] Kumar K, Desai V, Cheng L, Khitrov M, Grover D, Satya RV, et al. AGeS: a software system for microbial genome sequence annotation. PLoS One 2011;6:e17469.

[18] Christian W, Esquembre F, Barbato L. Open source physics. Science 2011;334:1077–8.

[19] Marzullo TC, Gage GJ. The SpikerBox: a low cost, open-source bioamplifier for increasing public participation in neuroscience inquiry. PLoS One 2012;7:e30837.

[20] Pearce JM. The case for open source appropriate technology. Environ, Dev Sustainability 2012;14:425–31.

[21] Stokstad E. Open-source ecology takes root across the world. Science 2011;334:308–9.

[22] Bruns B. Open sourcing nanotechnology research and development: issues and opportunities. Nanotechnology 2001;12:198–210.

[23] Mushtaq U, Pearce JM. Open source appropriate nanotechnology. In: Maclurcan D, Radywyl N, editors. Nanotechnology and global sustainability; 2012. p. 191–213.

[24] Nielsen M. Reinventing discovery: the new era of networked science. Princeton University Press; 2011.

[25] Lang T. Advancing global health research through digital technology and sharing data. Science 2011;331:714–7.

[26] Glynn LH, Hallgren KA, Houck JM, Moyers TB. Cacti: free, open-source software for the sequential coding of behavioral interactions. PLoS One 2012;7:e39740.

[27] Denkenberger DC, Brandemuehl MJ, Pearce JM, Zhai J. Expanded microchannel heat exchanger: design, fabrication and preliminary experimental test. Proc Inst Mech Eng, Part A 2012;226:532–44.

[28] Denkenberger DC, Pearce JM. Compound parabolic concentrators for solar water heat pasteurization: numerical simulation. Proceedings of the 2006 international conference of solar cooking and food processing; 2006. p. 108.

[29] Brasor P. Public wary of official optimism. The Japan Times; Sunday, March 11, 2012.

[30] You Can View Official EPA Radiation Readings. Washington post. Available from: http://www.washingtonsblog.com/2011/03/you-can-view-official-epa-radiation-readings.html.

[31] Japan Geigermap: At-a-glance. Available from: http://japan.failedrobot.com/.

[32] Pearce JM. Limitations of nuclear power as a sustainable energy source. Sustainability 2012;4(6):1173–87.

[33] Burke J, Estrin D, Hansen M, Parker A, Ramanathan N, Reddy S, et al. Participatory sensing. Proceedings of the workshop on the world-sensor-web (WSW'06), mobile device centric sensor networks and applications; 2006. p. 117–34; Boulder, Colorado.

[34] Crooks A, Croitoru A, Stefanidis A, Radzikowski J. Earthquake: twitter as a distributed sensor system. Trans GIS 2013;17(1):124–47.

[35] Settle FA. Topics in chemical instrumentation, evolution of instrumentation for UV-visible spectrophotometry. J Chem Educ 1986:A216–23.

[36] Popov-Raljić JV, Mastilović JS, Laličić-Petronijević JG, Popov VS. Investigations of bread production with postponed staling applying instrumental measurements of bread crumb color. Sensors 2009;9:8613–23.

[37] Popov-Raljić JV, Laličić-Petronijević JG. Sensory properties and color measurements of dietary chocolates with different compositions during storage for up to 360 days. Sensors 2009;9:1996–2016.

[38] Popov-Raljić JV, Lakić NS, Laličić-Petronijević JG, Barać MB, Sikimić VM. Color changes of UHT milk during storage. Sensors 2008;8:5961–74.

[39] Bogue R. Optical chemical sensors for industrial applications. Sensor Rev 2007;27(2):86–90.

[40] Terry PA. Application of ozone and oxygen to reduce chemical oxygen demand and hydrogen sulfide from recovered paper processing plant. Int J Chem Eng 2010. http://dx.doi.org/10.1155/2010/250235; Article ID 250235, 6 pages.

[41] Elliot AM. A photoelectric colorimeter for estimating protozoan population densities. Trans Am Microscopial Soc 1949;68:228–33.

[42] Eckhardt L, Mayer JA, Creech L, Johnston MR, et al. Assessing children's ultraviolet radiation exposure: the potential usefulness of a colorimeter. Am J Public Health 1996;86:1802–4.

[43] Trujillo O, Vanezis P, Cermignani M. Photometric assessment of skin colour and lightness using a tristimulus colorimeter: reliability of inter and intra-investigator observations in healthy adult volunteers. Forensic Sci Int 1996;81:1–10.

[44] Pesce S. Use of water quality indices to verify the impact of Córdoba City (Argentina) on Suquía River. Water Res 2000;34:2915–26.

[45] Hur J, Lee BM, Lee TH, Park DH. Estimation of biological oxygen demand and chemical oxygen demand for combined sewer systems using synchronous fluorescence spectra. Sensors 2010;10:2460–71.

[46] Hur J, Cho J. Prediction of BOD, COD, and total nitrogen concentrations in a typical urban river using a fluorescence excitation-emission matrix with PARAFAC and UV absorption indices. Sensors 2012;12:972–86.

[47] Anzalone GC, Glover AG, Pearce JM. Open-source colorimeter. Sensors 2013;13(4):5338–46. http://dx.doi.org/10.3390/s130405338.

[48] United States Environmental Protection Agency. Methods for chemical analysis of water and wastes. US EPA; 1983.

[49] Gonzales J. Wastewater treatment in the fishery industry. Food and Agriculture Organization of the United Nations; 1995.

[50] Hach Company. Oxygen demand, chemical, method 8000 for water, wastewater and seawater. Available from: http://www.hach.com/asset-get.download.jsa?Id=7639983640.

[51] Pearce JM, Grafman L, Colledge T, Ryan L. Leveraging information technology, social entrepreneurship and global collaboration for just sustainable development. Proceedings of the twelfth annual national collegiate inventors and innovators alliance conference; 2008. p. 201–10.

[52] Buitenhuis AJ, Zelenika I, Pearce JM. Open design-based strategies to enhance appropriate technology development. Proceedings of the fourteenth annual national collegiate inventors and innovators alliance conference: open; March 25–27, 2010. p. 1–12.

[53] Pearce JM, Albritton S, Grant G, Steed G, Zelenika I. A new model for enabling innovation in appropriate technology for sustainable development. Sustainability: Sci, Pract Policy 2012;8(2):42–53.

[54] Warren J. Mobile (Android) version of spectral workbench. The Public Laboratory; 2012, Available from: http://publiclaboratory.org/notes/warren/6-12-2012/mobile-android-version-spectral-workbench, [accessed 19.10.12].

[55] Williams A, Gibb A, Weekly D. Research with a hacker ethos: what DIY means for tangible interaction research. Interactions 2012;19:14.

[56] Buechley L, Hill BM. LilyPad in the wild: how hardware's long tail is supporting new engineering and design communities. MIT; 2010.

[57] President's Commission on the Accident at Three Mile Island. The need for change, the legacy of TMI: report of the President's commission on the accident at three Mile Island. Washington, DC, USA: President's Commission; 1979.

[58] Thompson J, Thompson R, Bear D. TMI assessment, part 2 1995; Available from: [accessed 11.04.12] http://www.southernstudies.org/images/sitepieces/ThompsonTMIassessment.pdf.

[59] Gundersen A. Three myths of the Three Mile Island accident, lecture. Burlington, Vermont, USA: Fairewinds Energy Education Corp; 2009.

[60] Hatch MC, Beyea J, Nieves JW, Susser M. Cancer near the Three Mile Island nuclear plant: radiation emissions. Am J Epidemiol 1990;132(3):397–417.

[61] Wing S, Richardson D, Armstrong D, Crawford-Brown D. A reevaluation of cancer incidence near the Three Mile Island nuclear plant: the collision of evidence and assumptions. Environ Health Perspect 1997;105:52–7.

[62] Hatch MC, Wallenstein S, Beyea J, Nieves JW, Susser M. Cancer rates after the Three Mile Island nuclear accident and proximity of residence to the plant. Am J Epidemiol 1991;81(6):719–24.

[63] Gur D, Good WF, Tokuhata GK, Goldhaber MK, Rosen JC, Rao GR. Radiation dose assignment to individuals residing near the Three Mile Island nuclear station. Proc PA Acad Sci 1983;57:99–102.

[64] Talbott EO, Youk AO, McHugh KP, Shire JD, Zhang A, Murphy BP, et al. Mortality among the residents of the Three Mile Island accident area: 1979–1992. Environ Health Perspect 2000;108:545–52.

[65] Wing S, Richardson D. Collision of evidence and assumptions: TMI déjà view. Environ Health Perspect 2000;108:A546–7.

[66] Talbott EO, Zhang A, Youk AO, McHugh-Pemu KP, Zborowski JV. Re: "Collision of evidence and assumptions: TMI déjà view." Environ Health Perspect 2000;108:A547–9.

CHAPTER 7

The Future of Open-Source Hardware and Science

7.1. INTRODUCTION TO THE FUTURE

We are all somewhat accustomed now to the rapid advancement in technology and the younger (or at least younger at heart) we are, the more likely it is that we are at home with embracing the accelerated changes of bringing ever-greater technologies into our lives. In experimental science and engineering, the tools to participate in the advancement of this technology have skewed the costs so high that they are beyond the grasp of all but the most well-funded research groups. The cost of equipment and the cost of access to the literature have provided some form of brake to slow the progress. These brakes are wearing out and technological evolution is picking up speed. I was fortunate to enter into the scientific community just as the Internet was taking off and enjoyed the flood of knowledge that has driven the recent technological revolution. Today, my children not only have this unprecedented access to pure information and learning tools[1] in their home, but also with an evermore sophisticated version of the RepRap 3-D printers, they have the capability to fabricate research-grade scientific tools in the house. If they want a microscope to go look for tardigrades in the back yard, we can stop by Wal-Greens to pick up a few free, disposable camera lenses and then print the open-source microscope shown in Figure 7.1 or upgrade it to a digital microscope with your smartphone as seen in Figure 7.2 (thanks to the designs by Dr. Walus from Chapter 6).[2] Today, there are hundreds of such open-source tools available at costs affordable to most of the curious who did not previously have access. By the time my children reach high school, it appears clear that this collection will have expanded to cover all the research tools we are familiar with now, as well as a host of yet-undreamed wonders to probe the mysteries of the universe, our planet and even our bodies.

---

[1]For example, my 5-year-old daughter is learning programming so fast that it is a little scary. She is coding animated stories using Scratch, an open-source educational programming language and multimedia authoring tool released under GPLv2 license and Scratch Source Code License. http://scratch.mit.edu/.

[2]A fully printable microscope http://www.thingiverse.com/thing:77450. A printable microscope smartphone adapter http://www.thingiverse.com/thing:92355.

Open-Source Lab. http://dx.doi.org/10.1016/B978-0-12-410462-4.00007-X

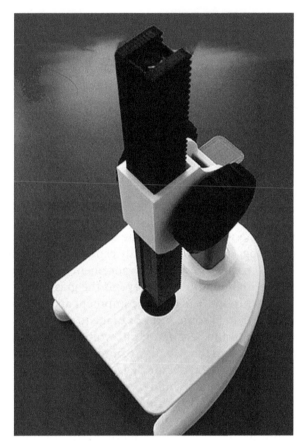

FIGURE 7.1
3-D printable open-source microscope.

7.2.   THE IMPACT ON THE SCIENTIFIC BRAIN DRAIN/GAIN

This advancement in the scientific tool arena has the potential to bring more people into the experimental and applied sciences. These people will come from what could be a reigniting of interests in the science, technology, engineering and mathematics (STEM) fields here in the United States and the rest of the west, but also a complete rearrangement of the migration routes of scientists from the developing world. Consider the current situation. I have taught at four universities in both the United States and Canada in a wide range of disciplines covering physics, materials science, mechanical engineering, computer science and electrical engineering—and in all of them, the vast majority of the graduate students were from other countries. The institutions I have worked for are

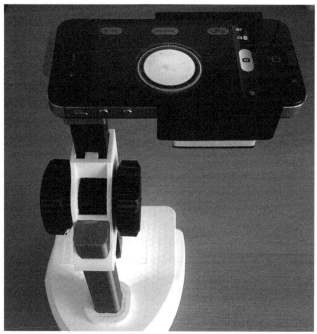

FIGURE 7.2
Upgraded 3-D printable open-source microscope to digital microscope with a printed component and a smartphone.

not particularly abnormal. For example, in 2006, the National Science Foundation reported that foreign students earned approximately 36% of the doctorate degrees in sciences and approximately 64% of the doctorate degrees in engineering [1,2]. Of these international students earning higher degrees in the sciences, about half of them stay in the United States, but this varies widely depending on the field and discipline (also the year and country) [3–5]: 64% for physical sciences, 63% for life sciences, 57% for mathematics, 63% for computer sciences, but only 38% in agricultural sciences. For some countries like China and India, the so-called stay rates can be substantially greater (88–92% on the high end). Obviously, having many of your best and brightest students come to the United States and stay here can have a large negative impact on the home countries [3–5]. On the other side of the coin, the U.S. economy and our supply of highly skilled and trained scientific personnel is benefited to an enormous degree.

There has been a long-standing concern about this situation as changes in the U.S. job market have made careers in science and engineering less attractive to Americans [1,6]. Science and engineering is perceived by many students as "hard" as compared to other disciplines such as business. Thus, the investment in overcoming the inherent challenge of the science and engineering disciplines does not appear worth it to the student when there appears to be more jobs in other fields. Historically, there have been, however, sufficient rewards to attract large numbers of "scientific immigrants" from the developing world. But what if this immigration stops (or significantly diminishes)? There are literally millions of exceptionally intelligent and driven young scientists who are growing up in the developing world today. The labs they have access to in their home countries are almost universally underfunded, which creates significant hurdles to participating in the experimental sciences. Without access to research-grade equipment, these scientists have generally three choices: (1) switching their primary focus on spreading information as teachers rather than creating new knowledge, (2) becoming theoreticians, and (3) moving to another country and doing experimental science.[3] To date, we have benefited enormously from many of them choosing option 3; however, a flood of low-cost, high-quality, open-source scientific hardware could accelerate a fourth choice—stay and help pull their home countries out of poverty. The U.S. share of the world's science and engineering graduates has been steadily declining as European and Asian universities, particularly those from China, have increased science and engineering degrees while U.S. degree production has remained more or less constant [6]. When these existing degree earners either start returning to their home countries in greater numbers or begin simply staying at home, the American dominance in science and engineering will erode further. This would result in our comparative advantage in the high-tech sector dying as well, with dire consequences for the American economy and more importantly, for the American worker. It may be perhaps tempting to then try to restrict open-source hardware for the continued stagnation of the status quo, but that is a losing battle. Consider, for example, that China already has a national open-source operating system based on Linux.[4] Where is the open-source U.S. operating system? If the United States does not embrace the open-source paradigm, we run the risk of our scientific education prominence and technological dominance following the death spiral of the Microsoft server market share into the inconsequential. Instead, it seems clear that we in the United States should aggressively capitalize on the opportunity to embrace the open-source paradigm and remain internationally competitive. If we do not, we will be overtaken by the accelerating innovation brought fourth by applying the open-source paradigm to scientific tools, which in turn accelerates all the other technologies. This last point can hardly be understated. Most working scientists are familiar with the beneficial effect of using a high-quality research tool after using antiquated equipment. It radically increases progress on individual projects. With the open-source paradigm offering nearly

---

[3]I do not mean to insinuate that choices 1 and 2 are in anyway less honorable or important than choice 3—some of my best friends are teachers and theoreticians.

[4]China, Canonical to collaborate on Ubuntu-based national OS http://www.extremetech.com/extreme/151381-china-canonical-to-collaborate-on-ubuntu-based-national-os. Even the mascot (a kylin – Google it) for the OS is inspiring.

universal access to an expanding array of high-grade tools, these projects in turn lead to faster development of applied technologies.

7.3. ACCELERATION OF TECHNOLOGICAL EVOLUTION

With the coming widespread access to inexpensive, high-quality, open-source scientific instruments, other technologies will move forward at much faster rates than to which we are accustomed. Imagine how fast science and technology will evolve when a large group of collaborators work together for mutual benefit. Consider even the myriad applications of the smartphone of today shown in Table 7.1[5], which is able to tackle a wide variety of tasks that in the past were divided between many separate devices. It is reasonable to assume that in the very near future, many devices we currently use could be incorporated into inexpensive and universally available smartphone devices. These advanced devices can act as scientific and engineering tools (such as the tricorder discussed in the last chapter) that will be available to everyone. For example, smartphone technology has already been developed as a tool to implement building energy audit programs to increase energy conservation measure (ECM) uptake and concomitant environmental and economic benefits [7]. This particular smartphone application provides an energy-audit platform with (1) quasi-real-time analysis, (2) continuous user engagement, (3) geospatial customization, (4) additional ECMs, (5) ECM ranking and user education, and (6) the ability to constantly evolve. My research group investigated a case study of such functionality that showed that distributed analysis with the use of a smartphone for 157,000 homes resulted in more than 50 years in savings to complete energy audits for all dwellings in the region following a traditional energy-audit model [7]. That is faster! This is just one application that could provide significant economic and environmental benefits with today's technology. But what of tomorrow's technology that is thrown forward with the power of the open-source paradigm?

7.4. OPEN-SOURCE RESEARCH IN THE FUTURE

As we have seen in detail in Chapters 4–6 and the literature [8–11] results show, the open-source hardware design approach develops extremely low-cost, high-quality, customized scientific instruments. The once onerous learning curve associated with open source has largely been overcome due to innovation

---

[5]This table was generated by several classes of Queen's University and Michigan Tech students to investigate the opportunity that smartphones provide to reduce individual ecological footprints with consolidating device functions into one—http://www.appropedia.org/Hardware_replaced_by_a_smartphone.

Table 7.1 Applications of Smartphones

Hardware Replaced with a Smartphone	Engineering Tools Replaced with a Smartphone
Land-line telephone	Flash light
Camera	Levels
Clock/watch/alarm clock	Graphing calculator
Calendar	Solar calculator
Boombox, mp3 player, CD or tape player	Light meter
DVD player	Handbook for engineering
TV	Decibel meter
Address book/rolodex	Data acquisition
Calculator	Image recognition
Voice recorder	Spectral light meter
Translators	Compass
GPS	Tape measure
Notebook	Telescope
Banking	GPS for coordinates of large spaces
Flash drive	Taper measurer for small things — length of phone
Remote control	Bluetooth comm — use to triangulate
Radio	Pedometer
Ebook reader	Two phones — sound — distance app
Computer (web browser)	One phone — length measurement with — scale
Library: books, encyclopedia, manuals, journals,	Database questioning
papers, monographs, magazines and newspapers	Scale
Mail	Throwing velocity, slapshot velocity
Games/game consol	Speedometer
Police scanner	Measure pressure, temperature — add tools
Travel tickets	Walkie talkie
Credit cards	Radiation detector
ID	
Banking/ATM	
Gambling cards	
Keys for car	

and rapid development of tools such as the Arduino prototyping platform discussed in Chapter 4, the RepRap discussed in Chapter 5, and associated software from the Linux community. The entire process of designing, printing and assembling new scientific research tools enables scientists to draw extensively on previously completed open-source work, requiring only a moderate literature review and moderate skill levels to improve or customize the tools and build. As more scientists take up this methodology and share back with the open-source scientific community, the time for another group to build their own high-quality instruments will continue to decline along with the cost, as the quality increases. These open-source tools can also be used as building blocks in tangential disciplines. So for example, the time for other research groups to create an open-source colorimeter following the details in Chapter 6 is also reduced for other applications beyond COD such as: measuring the concentration of some chemicals in a solution, quantifying observations of biological specimens, growth cultures, food science, quality control in manufacturing, diagnosis of diseases, testing the concentration of hemoglobin in blood, determining the efficacy of sun protection products, nephelometry for water quality, visibility, and global warming studies to measure global radiation balance, and many other applications [12–22]. The OpenSCAD code has been made available, making it easy to redesign the case to test, for example, alternative sizes or geometries of vials. In the same way, the Arduino software is easily altered, for example, to adjust integration time, light intensity or sensor sensitivity for another application.

For the example of the colorimeter, more work is necessary for this design to realize the full potential of the tool for all the applications discussed above. As scientists need these functionalities, they will build them and hopefully share them to provide the positive-feedback loop that will make it easier to do the next application. In addition, there is the potential to improve the functioning of the open-source colorimeter by making it portable, such as the incorporation of batteries and solar photovoltaic power. A multicompartment design[6] that can run more than one type of experiment using the same Arduino and control logic and simply adding additional LEDs and inexpensive sensors for different types of tests is already being developed and should be available when this book goes to press. Open-source wireless communication devices exist, making possible wireless communication between the instrument and a smartphone or tablet, thus augmenting data recording and analysis capabilities. Eventually, the capabilities can simply be built into a custom case for the smartphone or become integral to it. Modification of the OpenSCAD design may also permit the device to be used in-line for process control or quality assurance/quality control in industry itself. Again, this is but one small example, when in all likelihood, there will be thousands of such research tools developed in the near future.

As additional research groups begin to freely share the designs of their own open-source research tools, not only can the greater scientific community enjoy the same discounts on equipment, but following the FOSS approach, the equipment will evolve, becoming technically more advanced, easier to use and more useful. It is also likely that the price pressure from the open-source community [23] will drive down costs of commercial versions of the equipment, resulting in a decrease in overall research costs for everyone, even those that continue to rely on proprietary devices. For example, as discussed in Chapter 5, rapid advancement in 3-D printing technology has already produced plug-and-play 3-D printers at a price point equivalent to or lower than the cost of a good

---

[6]Customizer Colorimeter/Nephelometer Case v0.1 http://www.thingiverse.com/thing:47491.

computer. These low prices made available by the avoidance of intellectual property lock down discussed in Chapter 2 has created an explosion of 3-D printer companies, that currently number approximately a hundred companies.

The expense of sophisticated research-related equipment and tools has often limited their adoption to a select, well-funded few. This book has provided a methodology for applying the free and open-source hardware approach to the design and development of scientific equipment, a methodology that eliminates cost as a barrier to adoption and makes the tools available to the broadest possible audience. The performance of research equipment produced by this methodology has been successfully demonstrated against much more costly commercial products in a wide range of disciplines from biology to physics and chemistry to environmental science. With the digital designs of scientific instrumentation shared freely in the scientific community, old challenges like reproducibility of experiments should fade into memory. In the near future, if you want to replicate a really good experiment that you freely read about in *PLOS One* to build off it, you will download the necessary files for the equipment, use your 3-D printer to fabricate them and run the experiment with the settings that were in the attached files. Then you will get to the next step faster. Others will be doing the same and you can all collaborate and pool your data to accelerate scientific development at rates never even dreamed about in the past.

7.5. CONCLUDING THOUGHTS

The outlook for development of scientific-grade instrumentation utilizing the free and open-source hardware approach is extremely promising. Inexpensive open-source 3-D printers and free software have put one-off production of highly specialized tools within the grasp of the end user, bypassing historically expensive design, marketing and manufacturing steps. Perhaps more importantly, these technologies and methodologies promise previously unheard-of access to sophisticated instrumentation by those most in need of it, laboratories in underdeveloped and developing countries. Science will be helped and accelerated directly as well as indirectly, by reducing the costs of high-quality lab-based hands-on science and technology education. The future is bright as a virtuous cycle is created with the benefits of sharing designs helping to further accelerate research within your own laboratory. Join us and enjoy the ride.

REFERENCES

[1] Matthews CM. Foreign science and engineering presence in US institutions and the labor force 2008.

[2] National Science Foundation, Science and engineering indicators 2004, vol. 1, NSB04-01, Arlington, VA, January 15, 2004, pp. 3.38–3.39.

[3] Finn MG. Stay rates of foreign doctorate recipients from US universities, 2001. Oak Ridge, TN: Oak Ridge Institute for Science and Education; 2003.

[4] Finn MG. Stay rates of foreign doctorate recipients from US universities, 2003. Oak Ridge, TN: Oak Ridge Institute for Science and Education; 2005.

[5] Finn MG. Stay rates of foreign doctorate recipients from U.S. universities, 2005. Science and Engineering Education Oak Ridge Institute for Science and Education; 2007.

[6] Freeman RB. Does globalization of the scientific/engineering workforce threaten US economic leadership? (No. w11457) National Bureau of Economic Research; 2005.

[7] Leslie P, Pearce J, Harrap R, Daniel S. The application of smartphone technology to economic and environmental analysis of building energy conservation strategies. Int J Sustainable Energy 2012;31(5):295–311.

[8] Pearce JM. Open source research in sustainability. Sustainability: J Rec 2012;5(4):238–43.

[9] Pearce JM. Building research equipment with free, open-source hardware. Science 2012;337(6100):1303–4.

[10] Zhang C, Anzalone NC, Faria RP, Pearce JM. Open-source 3D-printable optics equipment. PLoS ONE 2013;8(3):e59840. http://dx.doi.org/10.1371/journal.pone.0059840.

[11] Anzalone GC, Glover AG, Pearce JM. Open-source colorimeter. Sensors 2013;13(4): 5338–46. http://dx.doi.org/10.3390/s130405338.

[12] Popov-Raljić JV, Mastilović JS, Laličić-Petronijević JG, Popov VS. Investigations of bread production with postponed staling applying instrumental measurements of bread crumb color. Sensors 2009;9:8613–23.

[13] Popov-Raljić JV, Laličić-Petronijević JG. Sensory properties and color measurements of dietary chocolates with different compositions during storage for up to 360 days. Sensors 2009;9:1996–2016.

[14] Popov-Raljić JV, Lakić NS, Laličić-Petronijević JG, Barać MB, Sikimić VM. Color changes of UHT milk during storage. Sensors 2008;8:5961–74.

[15] Bogue R. Optical chemical sensors for industrial applications. Sens Rev 2007;27:86–90.

[16] Terry PA. Application of ozone and oxygen to reduce chemical oxygen demand and hydrogen sulfide from recovered paper processing plant. Int J Chem Eng 2010. http://dx.doi.org/10.1155/2010/250235.

[17] Elliot AM. A photoelectric colorimeter for estimating protozoan population densities. Trans Amer Microsc Soc 1949;68:228–33.

[18] Eckhardt L, Mayer JA, Creech L, Johnston MR. Assessing children's ultraviolet radiation exposure: the potential usefulness of a colorimeter. Amer J Pub Health 1996;86:1802–4.

[19] Trujillo O, Vanezis P, Cermignani M. Photometric assessment of skin colour and lightness using a tristimulus colorimeter: reliability of inter and intra-investigator observations in healthy adult volunteers. Forensic Sci. Int. 1996;81:1–0.

[20] Pesce S. Use of water quality indices to verify the impact of Córdoba city (Argentina) on Suquía River. Water Res 2000;34:2915–26.

[21] Hur J, Lee BM, Lee TH, Park DH. Estimation of biological oxygen demand and chemical oxygen demand for combined sewer systems using synchronous fluorescence spectra. Sensors 2010;10:2460–71.

[22] Hur J, Cho J. Prediction of BOD, COD, and total nitrogen concentrations in a typical urban river using a fluorescence excitation-emission matrix with PARAFAC and UV absorption indices. Sensors 2012;12:972–86.

[23] Deek FP, McHugh JAM. Open source: technology and policy. Cambridge, UK: Cambridge University Press; 2007.

Index

Note: Page numbers followed by "f" denote figures; "t" tables.

A

ABS. *See* Acrylonitrile butadiene styrene (ABS)
Academic journals, open-source license for 41–42
Acrylonitrile butadiene styrene (ABS) 97–99, 138–139, 160–161, 170, 186–187, 189, 246–247, 249
Adey, A. 237–239
Aggressive academic sharing, advantages of 13–21
Agreement
 End User License Agreement 39
 shrink-wrap 39
altLab 45–47
Altruism 24–26, 30–31
Anticommons 37–38, 45–47, 52. *See also* Commons.
Anzalone, G. 218
Anzalone, J. 6f
Applie 156
Appropedia 15, 16, 19, 18–19, 22–23, 51, 99, 216–217
ArduCopter 25
Arduino 25, 59–62, 93, 204f, 209, 221, 223f, 227, 237–239
 board 60–62
 Due 64t–65t
 Geiger (radiation detector) 5–6
 -controlled open-source orbital shaker 8–9, 8f
 integrated development environment 154–156, 218
 LCD display, wiring diagrams for 78f
 Leonardo 64t–65t
 microcontroller 67–68
 platform 60, 186–187, 218

software 259–261
 Uno 61f, 64t–65t, 66f, 70, 178–181, 218
 working with 68–71, 68f
ArduPilot 25
 Mega 2 25
ArduPlane 25
Art of Illusion 165
arXiv 21–22
Asnaes Power Station 25–26
ATMEGA8 microcontroller 60
ATMEGA168 microcontroller 60
ATMEL 60
 ATmega328 170
AutoCAD 165
Automated pipetting system 59
Automation 59

B

Background material development, pre peer-review in 14–16
Bailey, S. 210
Banzi, M. 59–60
Beagle Bone 64t–65t
Beer–Lambert law 216, 219
Bessen, J. 14–16
Bill of materials (BOM) 51, 216–217, 227, 232
 by component 110–113
  extruder 113
  frame 110–111
  x-axis 111–112
  y-axis 112–113
  z-axis 111
 3-D printer 99, 100t–102t
Biology 226–241
 centrifuges 229–236
  DremelFuge 230–232, 230f–231f

microcentrifuge 232–233, 233f
 USB powered cytocentrifuge 233–236, 234f
 OpenPCR 226–229, 228f
 research tools 237–241
Blender (drawing package) 165
Blender CAD 165
BOM. *See* Bill of materials (BOM)
Bowden sheath 138–139, 141f
Bowyer, A. 95
BRL-CAD 165
BSD license 45
Buckner funnel 246–247, 246f
Budaschnozzle 139, 141f
Buildlog 2X Laser Cutter 189
BY-NC-ND license 45
BY-NC-SA license 45

C

C++ 156
CAD. *See* Computer-aided design (CAD)
CC. *See* Creative Commons (CC)
CC-BY license 46f
CC-BY-NC license 45, 46f
CC-BY-NC-ND license 46f
CC-BY-NC-SA license 46f, 213–214
CC-BY-ND license 45, 46f
CC-BY-SA license 45, 46f, 52, 220–221
Centrifuges 229–236
 DremelFuge 230–232, 230f–231f
 microcentrifuge 232–233, 233f
 USB powered cytocentrifuge 233–236, 234f
CERN OHL 38–39, 47, 242
Challenges to open-source research, overcoming 21–23

Charity 23–24
Chemical oxygen demand
    (COD) 216, 219–220
Chemistry 241–249
    spectrometers 241–246
    3-D printable chemical equip-
        ment 246–249, 246f–249f
Chromatography 188
Chronicle of Higher Education
    (Edmund Optics) 31–32
Chumby License 38–39, 44–45
Citations, open-source research
    impact on 18–19
COD. *See* Chemical oxygen demand
    (COD)
Collaboration 24–25, 30–31
Colorimeter 216–220, 217f, 219f,
    261
Commons 37–38
    anticommons 37–38, 45–47, 52
    intellectual 55
    tragedy of the 37. *See also* Creative
        Commons (CC).
Competition 25–26
Computer-aided design (CAD) 165,
    165–166
    AutoCAD 165
    Blender CAD 165
    BRL-CAD 165
    FreeCAD 165
    HeeksCAD 165
    Inventor 165
    NX 165
    OpenCASCADE 165
    OpenSCAD 9, 137, 165–169,
        169f, 171f, 173–174, 181–
        182, 185–187, 216–217, 217f,
        246–247, 259–261
    Python CAD 165
    Shapesmith 165
    Solid Works 165
    TinkerCAD 165
    Wings3D 165
Controller board
    installation 155–156
Cooperation 23–26
Copyleft 42–43
Copyright 40, 42–43
Copyright Act 53
Creative Commons (CC) 38–39
    license 45, 46f
        Arduino models under 59
        CC-BY 46f

CC-BY-NC 45, 46f
CC-BY-NC-ND 46f
CC-BY-NC-SA 46f
CC-BY-ND 45, 46f
CC-BY-SA 45, 46f, 52, 213–214,
    220–221
Crediting 37–38
Cuartielles, D. 59–60
Cura 154, 157, 158f, 167–169, 218
Customization 95, 160–161

D
Debian 48, 67, 155–156
Decentralization 14
Dichloromethane 246–247
DIY drones 25
DNA 226–227, 229
    extraction, portable cell lysis device
        for 6, 7f
Documentation 43
    open-source hardware definition
        1.0 48
    open-source license for 41–42
Dremel drill 6–7, 8f
DremelFuge 230–232, 230f–231f
    chuck 6–7, 8f
Dusseiller, M. 240–241

E
ECM. *See* Energy conservation mea-
    sure (ECM)
Electronics prototyping 59–60
Ellipsometers 188
End User License Agreement
    (EULA) 39
Energy conservation measure
    (ECM) 259
Engineering 189–215
    open-source laser welder 189–209
    radiation detection with
        open-source Geiger
        counters 209–213, 210f–214f,
        224–225
    Xoscillo 213–215, 215f
Environmental chamber 62–63
    Polar Bear. *See* Polar Bear Open-
        source Environmental
        Chamber
Environmental science 215–225
    colorimeter 216–220, 217f, 219f
    pH meter 220–225
EULA. *See* End User License Agree-
    ment (EULA)

Everywhere Tech 45–47
Experimental design
    development, pre peer-review
        in 14–16
    improved, sharing and 16–18
Experimental science 184
Extruded aluminum angle, 3-D
    printed optical mounts
    for 181–182, 181f
Extruder 133–139, 135f–143f

F
Facebook 2–3
Fairness 23–24
Faria, R. 72–75, 86–87
Firmware 153–154, 216–217
    configuration of 156–157
FLOSS. *See* Free libre open source
    software (FLOSS)
Forgiveness 23–24
FOSH. *See* Free and open-source
    hardware (FOSH)
FOSS. *See* Free and open-source
    software (FOSS)
Franchise model 30–31
Free access 14
Free and open-source hardware
    (FOSH) 5–9, 6f–9f, 25
    defined 5
Free and open-source software
    (FOSS) 2–3, 14, 25, 38–39,
    51
    defined 1–2
    development of 3–5
Free libre open source software
    (FLOSS) 1–2
Free software, defined 1–2
FreeCAD 165
Freeduino 221
Fritzing 218
Funding opportunities, increased,
    open-source research impact
    on 19–21
Future
    of open-source hardware 254–263
    open-source research in 259–262

G
Geiger counters
    Kit 209, 210f–211f
    radiation detection with 209–213,
        210f–214f

GitHub 51
GNU (GNU's Not Unix) 40
  licence 40–41
GNU Public License (GPL) 38–41,
    45
  comparison with Open Hardware
    License 43
GNU-GPL license 45
GNU/Linux. *See* Linux
Google 2–3, 19–21
  Analytics 19
  Sketchup 165
Governor, J. 40
GPL. *See* GNU Public License (GPL)
Gracie, A. 240–241

H
Hadoop, 3
Hackers 1–2
  ethic, for open-source
    research 14–16
Hackteria 240–241
Hardware
  fantastic and inexpensive scientific
    hardware, printing 160–161,
    160f
  free and open-source hardware
    5–9, 6f–9f, 25
  improved hardware design, sharing
    and 16–18
  Open Hardware License. *See* Open
    Hardware License (OHL)
  open-source. *See* Open-source
    hardware (OSH)
  vendors, fate of 31–32
HeeksCAD 165
High-performance liquid chromatog-
    raphy (HPLC) 248
h-index 19
HPLC. *See* High-performance liquid
    chromatography (HPLC)
Hydra GUI 203f–205f

I
IBM 40
IDE. *See* Integrated development
    environment (IDE)
Industrial strength sharing 25–31
Industrial symbiosis 25–31, 28f–29f
Inkscape 205f–206f
Integrated development environment
    (IDE) 60, 218
  Arduino 154–156

Intellectual commons 55
Intellectual monopoly
  rights, on commons software 40
  view, of open-source research 13
Intellectual property (IP) 13–15,
    21–22, 31–32, 37, 39
  continued challenges to 52–54
  obesity 45–47
  tragedy 53
Inventor 165
IP. *See* Intellectual property (IP)
*It Will Be Awesome If They Don't Screw
    It Up* (Weinberg) 54

J
Jansen, P. 17–18, 189

K
Kalundborg, Denmark 25–30

L
Laser cut test tube racks 237, 238f
Laser cutter
  Buildlog 2X Laser Cutter 189
  open-source 189, 190f
Laser welder 189–209, 191f–192f,
    193t, 194f–197f, 199f–209f
Lego blocks 186–187
LGPL license 45
Libre software. *See* Free software
License/licensing 37–38
  BSD license 45
  BY-NC-ND license 45
  BY-NC-SA license 45
  Chumby License 38–39, 44–45
  Creative Commons. *See* Creative
    Commons license
  GNU Public License 38–41, 43, 45
  GNU-GPL license 45
  LGPL license 45
  MIT license 45
  open. *See* Open Hardware License;
    Open-source licensing
  Simputer General Public
    License 38–39, 47. *See also*
    Patent(s/ing).
Limit switches 148, 149f–150f
Linux 2–3, 30–31, 60–62, 67, 155,
    157, 257–259
  Mint 67, 155
  origin and development of 2
  Ubuntu 2–3, 67
Lira 71

M
Macintosh OSX 60–62, 67
Makerbeam 231–232, 235–236,
    237f
Makerbot 24–25
Marlin 154, 156
Maskin, E. 14–16
Mass spectrometers 188
MAXIM 3323E pin assignment, for
    open-source polymer welding
    system 193, 196f
Melzi board 144, 145f, 152–153,
    155–157, 186–187
Melzi driver 154–155
Mendell Prusa 6f
MeshLab 246
Michelson interferometer 181–182,
    182f
Microcentrifuge 232–233, 233f
Microchannel heat exchanger
    fabrication, polymer laser
    welding system for 17–18,
    18f
Microcontrollers
  ATMEGA8 microcontroller 60
  ATMEGA168 microcontroller 60
  open-source 58–93, 170, 171f
    Arduino. *See* Arduino
    family 63–67, 64t–65t
    Polar Bear Open-source
      Environmental Chamber.
      See Polar Bear Open-source
      Environmental Chamber
Microsoft 2–3
Minix 2
MintDuino 64t–65t
MIT license 45
Monochromators 188
Mota, C. 45–47
Myexperiment 21
Myspectral 243–245

N
National Science
    Foundation 256–257
Natural selection 25–26
NEMA17 motor holder 189
Netduino 64t–65t
Netduino Plus 64t–65t
Neves, C. 221
96-well plate/0.2 ml strip tube
    magnet rack 237–239,
    239f

Nowak, M.
  Supercooperators—Beyond The
    Survival of the Fittest: Why
    Cooperation, not Competition, is
    the Key to Life 24
NX 165

O

ODS. *See* OpenDocument Spread-
  sheet (ODS)
OHL. *See* Open Hardware License
  (OHL)
"Open access +" 19
Open Hardware License (OHL)
  44–45, 47
  CERN 38–39, 47, 242
  Chumby License 38–39, 44–45
  comparison with GNU Public
    License 43
  Tucson Amateur Packet
    Radio 38–39
Open Solar Outdoors Test Field
  (OSOTF) 19–21
Open solar photovoltaic systems
  19–21, 25–31, 28f–29f
Open Source Hardware Association
  (OSHWA) 38–39, 47–50
  open-source hardware definition
    1.0 48–50
  open-source hardware statement of
    principles 1.0 48
  2012 survey 45–47
Open Source Initiative 41–42
OpenBeam system 171–182,
  231–232
  aluminum extrusion system 171,
    172f
  circular holder 174–175, 176f
  kinematic mirror or lens
    holder 174–177, 177f
  magnetic optics base 173–174,
    175f
  offset rod holder 173–174, 174f
  printed magnetic base 172–173,
    172f
  sample holder 174–177, 179f
  screen holder 174–175, 179f
  simple rod holder 173–174, 174f
  static fiber-optic holder 174–175,
    178f
  static filter or lens square 174–
    175, 176f
  T-brackets 172–173, 173f

OpenCASCADE 165
OpenDocument Spreadsheet
  (ODS) 51
openMaterials 45–47
Openness 14
OpenPCR (DNA analysis) 5,
  226–229, 228f
Openresearch 21
OpenSCAD 9, 137, 165–169, 169f,
  171f, 173–174, 181–182,
  185–187, 216–217, 217f,
  246–247, 259–261
  User Manual 167
Open-source, defined 1–5
Open-source hardware (OSH) 2, 188
  best practices for using 50–52
  definition 1.0 48–50
  etiquette for using 50–52
  future of 254–263
  licenses 38–39, 42–47
    Chumby License 38–39, 44–45
    using open-source software
      licenses of 45–47
    TAPR OHL 38–39, 42–44
  statement of principles 1.0 48
Open-source lab jack 160–161, 160f,
  175–177, 180f
Open-source licensing 36–57
  for academic journals 41–42
  classes, comparison of 41t
  continued IP challenges 52–54
  for documentation 41–42
  hardware licensing. *See* Open
    Hardware License; Open-
    source hardware
  software licensing
    fundamental requirements
      for 41
    of open-hardware, using 45–47.
      See also License/licensing.
Open-source microcontrollers
  58–93, 170, 171f
  Arduino. *See* Arduino
  family 63–67, 64t–65t
  Polar Bear Open-source
    Environmental Chamber.
    See Polar Bear Open-source
    Environmental Chamber. *See*
    also Microcontrollers.
Open-source optics 170–188,
  172f–180f
  component fabrication, material
    and energy costs associated
    with 182, 183t

education, open-source approach
  for, benefits of 184–186
future of 186–188
library 184–186
limitations of 186–188
research, open-source approach
  for, benefits of 181–184
Open-source parametric filter
  wheel 178–181, 180f
Open-source research
  aggressive academic sharing,
    advantages of 13–21
  background material development,
    pre peer-review in 14–16
  citations and 18–19
  experimental design,
    improved 16–18
  experimental design development,
    pre peer-review in 14–16
  funding opportunities,
    increased 19–21
  in future 259–262
  hardware design, improved 16–18
  overcoming challenges to 21–23
  public relations and 18–19
  student recruitment,
    increased 19–21
  visibility and 18–19
Open-source software
  comparison with proprietary
    software 40
  model 30–31
  as proprietary software 40
OpenWetWare 21
Optiboot 155
Optical chopper wheel 167–169, 169f
Optics, E.
  Chronicle of Higher
    Education 31–32
Orbital shaker 8–9, 8f
Oscilloscope
  Xoscillo 213–215
OSH. *See* Open-source hardware
  (OSH)
OSHWA. *See* Open Source Hardware
  Association (OSHWA)
OSOTF. *See* Open Solar Outdoors
  Test Field (OSOTF)

P

Page-view counting tool 18–19
Parametric automated filter wheel
  changer 9, 9f

Parametric cuvette/vial racks 5–6, 7f
"Participatory Sensing"
    framework 212–213
Partnership model 30–31
Patent(s/ing) 43
  law 53–54
  protection 14–15
  software 15–16
  thicket 54. *See also* License/licens-
    ing; Open Hardware License;
    Open-source licensing.
PCR. *See* Polymerase chain reaction
  (PCR)
Peer-review 37–38
  background material development,
    pre peer-review in 14–16
  experimental design development,
    pre peer-review in 14–16
Perens, B. 48
pH, defined 220–221
pH meter 220–225
pHduino 5, 221, 222f–226f,
  223–224
  Arduino shield details for 223f
  Datalogger 223–224, 225f–226f
  electronic schematic for 222f
  PCB layout of 222f
Physics
  education, open-source approach
    for, benefits of 184–186
  open-source optics 170–188,
    172f–180f
  research, open-source approach
    for, benefits of 181–184
Pipette stand 247, 247f
PLA. *See* Polylactic acid (PLA)
Platypus forceps 234–235, 236f
Polar Bear Open-source
    Environmental Chamber 5,
    71–93
  bill of materials for 75–77, 76t
  construction of 83–86, 84f–86f
  controller concepts 87–89, 88f
  controls of 86–87
  open-source technological evolu-
    tion for 92–93
  operation of 89–90, 89f–90f
  software 86–87
  specifications of 75, 75t
  troubleshooting of 90–92, 92f
  wiring diagrams 77–83, 78f–82f
Polyethylene 97–98
Polylactic acid (PLA) 97–98, 170

Polymer laser welding system, for
    microchannel heat exchanger
    fabrication 17–18, 18f
Polymerase chain reaction
    (PCR) 226–229, 228f
  tube racks 234–235, 235f
Polymers 190–191, 207
  acrylonitrile butadiene styrene
    97–99, 138–139, 160–161,
    170, 186–187, 189, 246–247,
    249
  laser welding system 191–192,
    191f–192f, 193t, 194f
  polylactic acid 170
Polyvinyl alcohol (PVA) 246–247
Portable cell lysis device, for DNA
    extraction 6, 7f
Printable adjustable volume
    pipette 248, 248f–249f
Printer interface 154, 157
Pronterface 154, 157
Proprietary software
  comparison with open-source
    software 40
  open-source software as 40
Prusa Mendel RepRap 98f, 99, 113
Public domain 40
Public Laboratory for Open Technol-
    ogy and Science 241–245
Public relations, information sharing
    and 18–19
PVA. *See* Polyvinyl alcohol (PVA)
Python CAD 165

R
Radiation detection with open-source
    Geiger counters 209–213,
    210f–214f, 224–225
Rand, A. 23–24
Rapid prototyping 95
Raspberry Pi 62, 64t–65t, 237–239
Recyclebots 181–182, 185–186
Rent seeking 39
Replicator 2 24–25
RepRap 5–6, 6f, 71–72, 83, 94–163,
    165–166, 169–170, 173–174,
    182, 185–189, 218, 231–233,
    237–239, 255
  building 99–153
    assembly instructions 113
    bill of materials by
      component 110–113
    electronics 142–145, 144f–148f

    extruder 133–139, 135f–143f
    frame 114–115, 114f–115f
    limit switches 148, 149f–150f
    power supplies 151–152,
      151f–152f
    x-axis 117–124, 119f–128f
    y-axis 115–117, 116f–118f
    y-stage, attaching 124–133,
      128f–130f, 132f–134f
    z-axis 117, 118f–119f
  family tree 96f–97f
  first time printing 158–161
    fantastic and inexpensive scien-
      tific hardware, printing
      160–161, 160f
    z-axis, leveling 158, 159f
  printed parts 103t–110t
  Prusa Mendel RepRap 98f, 99, 113
  software 153–157
    controller board
      installation 155–156
    firmware 153–154
    firmware configuration 156–157
    printer interface 154, 157
    slicer 154, 157
    sources of 154–155
  trim pots for motors,
    adjusting 152–153
Rights
  copyright 40, 42
  intellectual monopoly 40
  software 39–42

S
Sanguino 71
Scientific brain drain/gain,
    open-source impact
    on 256–259
Scientific hardware, fantastic and
    inexpensive, printing
    160–161, 160f
Secondary industry model 30–31
Seidle, N. 45–47
Selective laser sintering (SLS) 189
Selfishness 23–25, 37
SGPL. *See* Simputer General Public
    License (SGPL)
Shapesmith 165
Shapeways 230–231
Sharing
  aggressive academic sharing,
    advantages of 13–21
  benefits of 12–35

Sharing *Contiued*
  industrial strength 25–31
  reasons for 23–25
Shetty, Y. 240–241
Shrink-wrap agreement 39
Siderits, R. 234–236
Simputer 47
Simputer General Public License
    (SGPL) 38–39, 47
Skeinforge 167–169
Sketchbook 69–70
Sketches 69
Slic3r 167–169, 218
Slicer 154, 157
SLS. *See* Selective laser sintering (SLS)
Smartphones 259
  applications of 260t
Software
  Arduino 259–261
  free and open-source software
      2–3, 14, 25, 38–39, 51
  free libre open source
      software 1–2
  intellectual monopoly rights
      on 40
  libre. *See* Free software
  licensing
    fundamental requirements for 41
    of open-hardware, using 45–47
  open-source
    comparison with proprietary
        software 40
    model 30–31
    as proprietary software 40
  patenting 15–16
  Polar Bear Open-source Environ-
      mental Chamber 86–87
  proprietary
    comparison with open-source
        software 40
    open-source software as 40
  RepRap 153–157
    controller board
        installation 155–156
    firmware 153–154
    firmware configuration 156–157
    printer interface 154, 157
    slicer 154, 157
    sources of 154–155
  rights 39–42
  Spectral Workbench 241–242,
      243f

Solar water pasteurization
    system 190–191
Solid Works 165
SourceFourge.net 86
Spectral Workbench software
    241–242, 243f
Spectrometers 188, 241–246
  desktop spectrometry kit 242
  foldable 242
  smartphone 242
  v3.0 242, 244f
Spectruino 243–245, 244f
SQL Server 3
Stallman, R. 40
Star Trek 243–245
STereoLithography (STL) files 99,
    165–166, 181–182
STL. *See* STereoLithography (STL) files
Student recruitment, increased,
    open-source research impact
    on 19–21
*Supercooperators—Beyond The Survival
    of the Fittest: Why Cooperation,
    not Competition, is the Key to
    Life* (Martin Nowak) 24
Survival of the fittest, evolutionary
    theory of 23–24
Synamptic Package Manager 67

T

Taos TSL230R 218
TAPR. *See* Tucson Amateur Packet
    Radio (TAPR) OHL
Tardi-GelBox 240–241, 241f
Technological evolution, acceleration
    of 259
Thingiverse 5–6, 16, 17–18, 42,
    51, 154–155, 181–182, 189,
    216–217, 232, 237, 240–241,
    246–248
  Customizer App 165–167, 168f
3-D printable cassette racks
    234–235, 236f
3-D printable chemical equip-
    ment 246–249, 246f–249f
3-D printable digital
    microscope 257f
3D-printable Laser Cutter 17–18
3-D printable open-source
    microscope 256f
3-D printable rapid fluid filters
    234–235, 235f

3-D printable test tube racks 237,
    239f
3-D printers 169–170, 186, 188–189
  adapter 6, 7f
  bill of materials 5–6, 6f, 99,
      100t–102t
  DremelFuge chuck 6–7, 8f
  RepRap. *See* RepRap
  parametric cuvette/vial racks 5–6,
      7f
  as reactionware for chemical
      synthesis 188
  solar-powered 185–186
3-D printing 95, 178–184, 186–187,
    189
3-D rapid prototyping 187–188
TinkerCAD 165
TinyDuino 71
Tinytim 232
Tomasello, M. 24–25
Torvalds, L. 2
Tricorder 243–245, 245f
Tucson Amateur Packet Radio (TAPR)
    OHL 38–39, 42–44
Tymrak, B. 207–209

U

UAV. *See* Unmanned aerial vehicle
    (UAV)
Ubuntu 2–3, 67
  Synamptic Package Manager 67.
      See also Linux.
UNIX 40
Unmanned aerial vehicle (UAV) 25
USB powered cytocentrifuge
    233–236, 234f

V

Vandalism 22–23
Visibility of academic work, informa-
    tion sharing and 18–19

W

Walus, K. 248–249
Web 2.0 2–3, 15–16
Weinberg, M.
  *It Will Be Awesome If They Don't
      Screw It Up* 54
Wikipedia 15–16
  infographic by Statista 2–3, 4f
Windows 60–62
  Azure, 3

Wings3D 165
Wiring diagrams (Polar Bear
        Open-source Environmental
        Chamber) 77–83
    for Arduino LCD display
        78f

for fan 82f
for humidifier 81f
for humidity sensor 80f
for keypad 79f
for level shift circuit 80f
for power resistors 81f

    for refrigerator/cooler 82f
World improvement 14

X

Xoscillo 5, 213–215, 215f
X-ray diffraction systems 188

Printed and bound by CPI Group (UK) Ltd, Croydon, CR0 4YY

03/10/2024

01040321-0001